Numerical Geometry of Images

Springer
New York
Berlin
Heidelberg
Hong Kong
London
Milan
Paris
Tokyo

Ron Kimmel

Numerical Geometry of Images

Theory, Algorithms, and Applications

 Springer

Ron Kimmel
Department of Computer Science
Technion—Israel Institute of Technology
Haifa 32000
Israel
ron@cs.technion.ac.il
www.cs.technion.ac.il/~ron

Cover: Illustration by Michael and Alex Bronstein.

Library of Congress Cataloging-in-Publication Data
Kimmel, Ron.
    Numerical geometry of images : theory, algorithms, and applications / Ron Kimmel.
      p. cm.
    Includes bibliographical references and index.
    ISBN 0-387-95562-3 (acid-free paper)
    1. Geometry, Differential—Data processing. I. Title.
    QA641.K56 2003
    516.3′6—dc21   20003052963

ISBN 0-387-95562-3   Printed on acid-free paper.

Printed in the United States of America.

9 8 7 6 5 4 3 2 1   SPIN 10889579

www.springer-ny.com

Springer-Verlag New York Berlin Heidelberg
A member of BertelsmannSpringer Science+Business Media GmbH

To my family

Preface

Perception of visual information occupies about 70% of our cortex activity. Apparently, about the same percentage of visual content occupies communication networks over the world. Thus, it is not surprising that as scientists we find it fascinating to process visual information and explore related fields in which we evaluate and model the formation of images and try to interpret their content. For example, we try to understand the process of visual perception, and invent new methods of analysis of visual data, such as our perception of color. What is a color image? And how can it be efficiently described, processed, and analyzed? In our study we need some sort of vocabulary and machinery to first formulate the problems that explain natural phenomena, and then find efficient solutions. This book provides a new view point on some classical problems, and uses classical tools to solve new problems. The variational geometric methods and resulting numerical schemes that we review in this book have already proven to be useful in various industrial applications where image analysis of complicated scenarios is involved.

Intended Use

This book is intended as a graduate textbook for engineers and computer science students. Unlike other monographs in this area, which assume deep prior knowledge, we provide the required mathematical basis using a sequence of exercises that lead the reader to an understanding of the tools we use. This is true for calculus of variations and differential geometry, as well as numerical analysis methods.

Prerequisites

We assume knowledge in basic calculus, linear algebra, and geometry. Beyond that, the basics are reviewed and taught through a sequence of exercises, with solutions to selected ones. These enable a smooth introduction of advanced methods and related applications.

Features

The book introduces modern and classical mathematical tools to the field of image analysis and processing. It has been taught as a one-semester course for graduate and undergraduate students at the Technion, Israel and as an invited tutorial in various forms by industrial companies and universities. The focus of the book is more on the tools rather than the problems, and in many cases problems from other disciplines are given as examples.

Roughly speaking, the book is divided into two main parts. The first five chapters are an introduction to the mathematical and numerical tools we use for the algorithms and applications that are presented in the second part of the book. The mathematical language we use is heavily influenced by differential geometry of curves and surfaces, as well as calculus of variations. Therefore, we start with brief introductions to these two powerful mathematical tools in Chapters 1 and 2. Next, Chapter 3 adds motion to the curves and surfaces and allows them to dynamically propagate. It also studies the properties of the evolving curves, such as curvature flow, also known as the "geometric heat equation," and affine flow. We can use dynamic curves as steepest descent processes of geometric models that we meet again later in the book. For example, the curvature flow is the result of arclength minimization of a planar curve, yet it also plays a major role in "total variation" minimization in image processing, as we meet again in Chapter 10. The link between propagating curves and dynamic images is the Osher–Sethian level set formulation introduced in Chapter 4. This modern formulation of curve evolution and general front propagation is a natural bridge between image processing and dynamics of embedded propagating curves. Chapter 5 brings some numerical considerations of working with the level set formulation.

Mathematical morphology is the main theme of Chapter 6. This field deals with set operations on geometric structures. In simple words, it is the algebra of shapes. We touch upon this field only through its connection to curve evolution, and relate to this aspect of the whole theory. If we restrict some of the shapes to be convex, then the basic morphology operations are closely related to the problem of computing distances. Efficient and accurate methods of computing distances in various geometric scenarios bring us into the fast marching algorithms presented in Chapter 7. Next, we are ready to explore the classic problem of shape from shading, which is presented in Chapter 8. The shape from shading is a toy problem in the field of computer vision by which we link various methods that we introduce and study in previous chapters. We show how this problem can be formulated as a solution of an eikonal equation and various classical and modern numerical solutions are reviewed.

After this exercise we tackle one of the major problems in image analysis, known as the "segmentation problem." Chapter 9 introduces a set of variational solutions to the segmentation problem in which geometry plays a dominant part. It shows, for example, that thresholding is a result of minimal variation criterion that can be coupled with other integral measures such as the "geodesic active contour" model. We move from curves to surfaces by introducing the concept of image as a surface in Chapter 10. The Beltrami framework considers images as surfaces embedded in the hybrid spatial-intensity space. This framework becomes interesting when considering color images as surfaces in a five-dimensional space: two dimensions for the location in the image and three dimensions that capture the geometric structure of our color perception. We show how to selectively smooth images while minimizing the area of the image surface. Finally, chapter 11 tackles one of the most challenging problems in image analysis, the face recognition problem. It reviews a new method I recently discovered and explored with my students Asi Elad and Michael and Alexander Bronstein. The bending invariant signature is defined and used to match isometric surfaces. It translates the problem of matching flexible surfaces into a simpler one of matching rigid objects. Equipped with such a powerful tool we are able to distinguish between identical twins under varying facial expressions based on a three-dimensional image of their facial surface. Each chapter includes graphic sketches, images, figures, and exercises at the end, with solutions to selected exercises in Chapter 12.

Intellectual Support

The book's Web Site includes slides for both students and instructors, as well as color illustrations. The solutions to selected exercises appear at the end of the book. New exercises will be added annually, at the course Web Site at: $http : //www.cs.technion.ac.il/ \sim ron/ng.html$.

Acknowledgment

I learned and helped to develop some of the numerical methods and geometrical algorithms covered in these lecture notes while collaborating with Alfred M. Bruckstein from the Technion, Israel, James A. Sethian from

UC Berkeley, Ravi Malladi from Berkeley Labs., David Adalsteinsson from the University of North Carolina at Chapel Hill, Laurent Cohen from Paris University Dauphine, Guillermo Sapiro from the University of Minnesota, Vicent Caselles from the University of Barcelona, Doron Shaked from HP Labs, Israel, and Nahum Kiryati and Nir Sochen from Tel-Aviv University. I am grateful to these brilliant researchers with whom I had the pleasure to collaborate. Nir Sochen, James Sethian, Alfred M. Bruckstein and Joachim Weickert gave valuable comments on early drafts.

I thank my graduate students Alexander Brook, Gil Zigelman, Alon Spira, Adi Liav and the students and the participants from the HP Laboratory and Applied Materials research group, of the "Numerical Geometry of Images" course held at the Technion during the winter semesters of 1999 through 2003, for their valuable comments and careful reading of the manuscript. I thank Yoav Schechner and Didi Sazbon from the Technion, for correcting typos in early versions of this manuscript. During the last year parts of the course were given as tutorials, short seminars, and invited lectures in Sydney, Australia, Copenhagen, Denmark, and all over Japan, France, and California. I thank all the participants who made valuable comments that helped me rewrite and refresh some of the chapters. I am grateful to Yana Katz, Hadas Heier, and Deborah Miller for their time and effort in LaTeXing and shaping my notes.

Feedback

Comments, suggestions, and corrections are most welcome by E-mail: ron@cs.technion.ac.il

Ron Kimmel
Technion—IIT, Haifa, Israel
July 2003

Contents

1

Introduction

The goals of this book are to introduce recent numerical methods motivated by differential geometry and to show how to apply them to various problems in image processing, computer graphics, robotic navigation, and computer vision. The mathematical machinery we build upon sits at a crossroad between many different disciplines like numerical analysis, differential and computational geometry, complexity analysis, topology and singularity theory, calculus of variation, and scale space theory.

Computer vision tries to answer questions like how should the computer interpret what the camera "sees." It starts from models of perceptual understanding and tries to develop algorithms that give an automatic interpretation of what is featured by an image. One example that we use in this book is the shape from shading problem, in which we try to reconstruct the three-dimensional shape of a smooth object from a single given gray-level image. To be more precise, we are interested in the 2D boundary manifold embedded in 3D space of the 3D object, that is, the boundary between our object and the rest of the world. Note that we introduced two new mathematical terms, *manifold*, that for the time being we regard as smooth mapping, and *embedding space*, which is the space in which our object lives. These are nonrigorous definitions that just give a geometric flavor.

In image processing we try to apply signal processing tools to enhance, sharpen, improve, denoise, and compress images. Obviously, a good model of our vision process would help us design better procedures. Next, computer graphics deals with numerical aspects of interpolation and approximation—how an object should be presented on the screen, and appear real to the human observer. How should the numerical data be transmitted to the viewer?

We see that these fields include phases of introducing images as an input to our computer algorithms, understanding, approximating, interpolating, or generally manipulating this data, generating synthetic data, and displaying the numerical data efficiently to the human observer; see Figure 1.1.

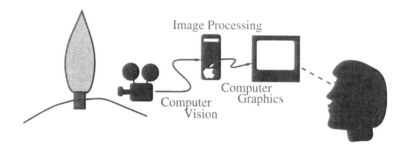

Figure 1.1: Computer vision, image processing, and computer graphics deal with input images, processing the data, and displaying the results.

A somewhat related field is robotic navigation, where we deal with questions like how should a robot move from point A to point B in a given dynamic or static environment. One should consider the geometry of the space in which the robot moves, and the properties of the robot.

1.1   Mathematical Tools and Machinery

We explore and use basic concepts in numerical analysis. One example is how to approximate a derivative of a function on a discrete grid and how to measure the error that we introduce by the approximation. Most of the problems we deal with have some geometric flavor, and thus we use classical tools from differential geometry that help us choose a coordinate system that simplifies the problem.

Differential geometry in general deals with the problem of finding invariant measures. We use curvature, normals, tangent planes, and other basic concepts in Riemannian differential geometry in our applications.

In calculus we know that when the derivative of a smooth function is zero there is an extremum point. That is, we search for the variation of the function along the argument axis that is equal to zero as an indication for a maximum, a minimum, or an inflection point. In a very similar way, calculus of variation is a powerful mathematical tool in the search for extremum configurations of functionals. By introducing a small variation to an integral of a function and its derivatives, we can come up with a set of conditions for the extremum configurations. These set of conditions is referred to as the *Euler–Lagrange* equations or, in short, EL equations.

For example, the measure of Euclidean arclength, $ds^2 = dx^2 + dy^2$, along a parametric curve $C(p) = \{x(p), y(p)\}$ is given by

$$\int_0^1 \left| \frac{\partial C(p)}{\partial p} \right| dp = \int_0^1 \sqrt{x_p^2 + y_p^2} dp.$$

We sometimes use the short-hand notation $C_p = \partial_p C = \partial C / \partial p$, and in case of one parameter we also use $C_p = C'$. Now, let us introduce a small variation to the arclength of a curve defined by the function $y(x)$, namely $C(x) = \{x, y(x)\}$ given by

$$L(y_x) = \int_{x_0}^{x_1} \left| \frac{\partial C(x)}{\partial x} \right| dx = \int_{x_0}^{x_1} \sqrt{1 + y_x^2} dx,$$

such that $y(x_0) = y_0$ and $y(x_1) = y_1$. Or in a more general notation, we say that $L(y, y') = \int F(y, y') dx$. Let $\tilde{y}(x) = y(x) + \epsilon \eta(x)$, where $\eta(x)$ is called the variation, and assume $\eta(x_0) = \eta(x_1) = 0$. See Figure 1.2. We take the derivative of $L(\tilde{y}(x), \tilde{y}'(x))$ with respect to ϵ, at $\epsilon = 0$, and check when it is equal to zero.

$$
\begin{aligned}
\frac{\delta L}{\delta y} \equiv \frac{dL(\tilde{y}, \tilde{y}')}{d\epsilon} \bigg|_{\epsilon=0} &= \frac{d}{d\epsilon} \int F(\tilde{y}, \tilde{y}') dx \\
&= \int \frac{d}{d\epsilon} F(\tilde{y}, \tilde{y}') dx \\
&= \int \left(\frac{\partial F}{\partial \tilde{y}} \frac{d\tilde{y}}{d\epsilon} + \frac{\partial F}{\partial \tilde{y}'} \frac{d\tilde{y}'}{d\epsilon} \right) dx \\
&= \int \left(\frac{\partial F}{\partial \tilde{y}} \eta + \frac{\partial F}{\partial \tilde{y}'} \eta' \right) dx \\
&= \int (\partial_{\tilde{y}} F) \eta \, dx + \eta \frac{\partial F}{\partial \tilde{y}'} \bigg|_{x_0}^{x_1} - \int \eta \frac{d}{dx} (\partial_{\tilde{y}'} F) dx \\
&= \int (\partial_{\tilde{y}} F) \eta \, dx - \int \eta \frac{d}{dx} (\partial_{\tilde{y}'} F) dx. \\
&= \int \left(\partial_{\tilde{y}} F - \frac{d}{dx} (\partial_{\tilde{y}'} F) \right) \eta(x) dx. \quad (1.1)
\end{aligned}
$$

In the above derivation we applied the chain rule, change of derivatives, followed by integration by parts, and the fact that η vanishes at the boundaries. Then, for $\frac{dL}{d\epsilon} \big|_{\epsilon=0} = 0$, to hold for every selection of $\eta(x)$, we need the following condition:

$$\partial_y F - \frac{d}{dx} (\partial_{y'} F) = 0,$$

known as the *Euler–Lagrange* equation. For our specific example where $F(y') = \sqrt{1 + y'^2}$, we have $\partial_{y'} F = y'/\sqrt{1 + y'^2}$ and $(d/dx)\partial_{y'} F = y''/(1 + y'^2)^{3/2}$. We will see later that this specific EL equation states that the curvature along the shortest curve connecting two points should be zero, and as we all know the shortest curve connecting two points is indeed a straight line with zero curvature, along which $y'' = 0$. We will revisit the curvature and introduce many other formulations and extensions for it.

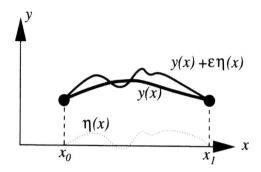

Figure 1.2: Adding the variation η to the curve $\{x, y(x)\}$.

1.2   Applications

What kind of practical problems can we solve with numerical methods enforced by differential geometry tools? Let us start with a brief review of curve evolution-based applications, and then add one dimension to the problem and explore surfaces and images that minimize their area as they flow toward a minimal surface.

Any gray-level image can be viewed as a set of its equi-intensity curves, or as a two-dimensional surface given by a function of the intensity I and the image plane coordinates (x, y), that is, $\{x, y, I(x, y)\}$. Several image analysis algorithms are based on propagating planar curves in the image plane according to local variations in the gray level of the image [6, 26, 93, 90, 176, 203, 204]. Those planar contours might be, for example, the level sets of an object whose shaded image we are trying to interpret so as to recover its three-dimensional structure. The shape from shading problem is a good example illustrating the way curve propagation algorithms found an interesting application [25]. Their usefulness in this and other applications was further enhanced by the recent development of an algorithm in the field of numerical analysis for the stable propagation of planar curves according to a variety of rules [161].

Other fields in which there were immediate consequences of having a stable and efficient way to propagate curves are computer aided design, robotics, shape analysis and computer graphics.

In CAD (computer aided design) there is a need to find offset curves and surfaces, implying constant-speed curve propagation. Geodesic deformable models were introduced for shape modeling and analysis. In computer graphics, Pnueli and Bruckstein found an interesting application in the design of a clever half-toning method they named **Digi$_D$ürer** [165, 166].

In robotics, where one often needs to plan a path for robots from a source point to a certain destination, one could determine shortest routes by propagating a wave of possibilities and finding out the way its wavefront reaches the destination point. This idea can be extended and used even in the presence of moving obstacles. We will mention the field of mathematical morphology, where there is a need to precisely compute various types of distance functions, to enable erosion or dilation of shapes.

The solutions to some of the problems that we explore are based on the ability to find a curve evolution-based formulation to the problem. This formulation is of the form of a differential equation that describes the propagation of a planar curve in time, under the constraints imposed by the problem. An evolving planar curve must often overcome various problems such as topological changes (e.g., a single curve that splits into two separate curves) and numerical difficulties that may be caused by the type of curve representation used (e.g., the problem of determining the offset curve to a polynomial parametric curve).

The most general propagation rule for a closed planar curve in time along its normal direction $\vec{N}$ is

$$\frac{\partial C}{\partial t} = V\vec{N} \quad \text{given} \quad C(0),$$

where $C(p,t) : S^1 \times [0,T) \to \mathbb{R}^2$ is the curve description, S^1 denotes the unit circle (sphere of dimension 1) that serves as a periodic parameterization interval for p, that is, a mapping from a circle to the simple closed curve. t is the time of evolution that starts from an initial given curve and evolves by V, a smooth scalar velocity in the normal to the curve direction $\vec{N}$. The function V may depend on local properties of the curve or on some external control variables like the image's gray level. V may also depend on the global properties of the curve (e.g., [182]).

1.2.1   Shape from Shading

The shape from shading problem is a classical problem in the area of computer vision [87, 89]. Its goal is to reconstruct a surface in 3D $\{x, y, z(x,y)\}$ from a single given gray-level picture $I(x,y)$. In [25, 92, 116, 101], it is shown that under reasonable assumptions about the light source and the object reflection properties, it is possible to solve this problem by using the image data to control the evolution of a planar curve so as to track the equal-height contours of the object. In this evolution rule the propagation time indicates the height with respect to the light source direction $\hat{l}$.

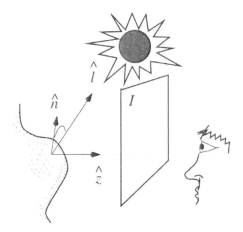

Figure 1.3: According to the simplest shading rule $I = \langle \hat{l}, \hat{n} \rangle$, where $\hat{n}$ is the surface normal.

The simplest *Lambertian* shading rule is given by

$$I(x,y) = \langle \hat{l}, \hat{n}(x,y) \rangle = \cos \angle(\hat{l}, \hat{n}),$$

where $I(x,y)$ is the shading image, the light source direction is $\hat{l}$, and $\hat{n}$ is the surface normal (see Figure 1.3). Consider the simplest case where the light source $\hat{l} = \{0,0,1\}$, that is the light source and the viewer are located at the same direction. The shading image in this case is given by $I(x,y) = 1/\sqrt{1 + z_x^2 + z_y^2}$, or equivalently $|\nabla z(x,y)| = \sqrt{1 - (I(x,y))^2}/I(x,y)$, and the evolution rule that takes us from one equal-height contour $C(0)$ to another $C(dt)$ is given by $C(dt) = C(0) - dt|\nabla z|^{-1}\vec{N}$, where $\vec{N}$ is the normal to the curve. In this derivation Bruckstein exploited the fact that the equal-height contour normal coincides with the gradient direction to solve the ambiguity in the surface normal while integrating the surface from its shading image.

The differential evolution of the equal-height contour, as determined by the gray-level image, is given by

$$C_t = \frac{I(x,y)}{\sqrt{1 - I(x,y)^2}}\vec{N}.$$

In [100], motivated by the height evolution equation, it is shown how to solve the global topology of the shape from shading problem for smooth surfaces (Morse functions).

1.2.2   Gridless Halftoning

Using the image data to control the evolution of a planar curve can be used to generate graphical effects. One such application for gray and color half-toning is the **Digi**$_D$*ürer*, which aims to emulate the work of classical engravers [165, 166, 184]. Figure 1.4 shows the result of propagating planar curves controlled by the image intensity.

A rough description of this evolution is

$$C_t = F(I(x,y))\vec{N},$$

where $F : \mathbb{R}^+ \to \mathbb{R}^+$ is a monotone function. Other image synthesis methods in computer graphics can be found in [158].

Figure 1.4: Result of the **Digi**$_D$*ürer* taken from [165].

1.2.3 Continuous-Scale Morphology

In the field of shape theory, it is often required to analyze a shape by applying some "morphological" operations that make use of a "structuring element" with some given shape. In [177] the problem of continuous-scale morphological operators is explored for cases in which the *structuring element* may be of any convex shape with variable sizes [21]; see also [3]. This problem too may be reformulated as the problem of applying a propagation rule for the shape boundary. The evolution rule for the shape's boundary is determined by the structuring element's shape $r(\theta)$, and the time of evolution in this case represents the size of the element. The planar evolution of the boundary curve is

$$C_t = \sup_{\theta} \langle r(\theta), \vec{N} \rangle \vec{N};$$

see Figure 1.5.

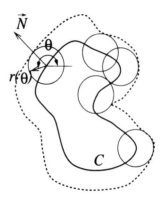

Figure 1.5: Dilation operation with a circle structuring element. In this case, since $\sup_{\theta} \langle r(\theta), \vec{N} \rangle = 1$, the evolution rule is simply $C_t = \vec{N}$, which is the offsetting or prairie-fire model.

1.2.4 Shape Offsets or Prairie-Fire Propagation

In CAD one often encounters the need to find the offset of a given curve or surface. A simple algorithm that solves this problem may be constructed by considering a curve that propagates with a constant velocity along its normal direction at each point [99, 17]. The propagation time represents the "offset distance" from the given curve, and the evolution rule is simply

$$C_t = \vec{N}.$$

This is, of course, also Blum's prairie-fire propagation model for finding shape skeleton, that is, the shock fronts of the propagation rule.

1.2.5   Minimal Geodesics on Surfaces

An important problem in the field of navigation is to find the shortest path connecting two points. It can be solved by considering an equal-distance contour propagating from a point on a given surface. In [97], an analytic model that describes the propagating curve embedded in the surface is introduced. Tracking such a curve in 3D is quite a complicated task. However, it is also possible to follow its projection on the coordinate plane.

Calculating distance maps on surfaces by tracking the projected evolution from both source and destination points on the given surface enables us to select the shortest path, which is given by the minimal level set in the sum of the two distance maps; see Figure 1.6.

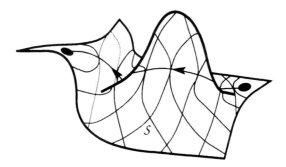

Figure 1.6: The minimal geodesic is the minimal set of the sum of the two geodesic distance maps.

The propagation time in this case indicates distance on the surface, namely the geodesic distance. This approach motivated a very efficient method for calculating distances on triangulated weighted domains recently introduced in [112].

1.2.6   Shortening Curves on Surfaces via Flow on a Plane

Given a path connecting two points on a given surface, it is sometimes required to shorten its length locally and to find the closest geodesic to the given curve. In [109] it is shown that this operation, too, may be done by propagating a curve along the geodesic curvature. This surface curve propagation may also be performed by tracking its planar projection onto the coordinate plane and may be used to refine minimal geodesics obtained by other methods, for example, the minimal path estimation obtained by the Kiryati–Szekely algorithm [121].

1.2.7 Distance Maps and Weighted Distance Transforms

As stated in [105], some of the above results may in fact be grouped under the same title of "generalized distance maps." While searching for offset curves, one constructs the distance transform. Reconstructing the shape from shading may be shown to be equivalent to calculating a weighted distance transform. Continuous-scale morphology may be shown to result in the distance transform under a given metric, where the structuring element of the morphological operations defines the unit sphere of the given metric.

1.2.8 Skeletons via Level Sets

"Skeletons are thin, exact descriptors of shapes" [195]. Define the distance of a point from a curve as the infimum of distances between the point to the set of curve points. The skeleton, or medial axis, of a shape is the set of internal points whose distance to the boundary is realized in more than one boundary point. Each skeleton point is associated with a width descriptor corresponding to its distance from the boundary (see Figure 1.7).

Figure 1.7: The skeleton of a connected planar shape is connected.

Being a stick figure, or naive description of the shape, skeletons are perceptually appealing. From a pattern recognition point of view, skeletons provide a unique combination of boundary and area information. Although analytically well defined in the continuous plane, there are numerical difficulties in extracting the skeletons. This situation has brought numerous solutions referred to as skeletonization or thinning algorithms.

A stable scheme for computing distances solves many of the inherent difficulties of skeletonization. As shown in [115], skeletons are located on zero crossing curves of differences of distance transforms from boundary segments. Applying simple differential geometry results to skeletons, it is possible to find a necessary and sufficient partition of the boundary to segments whose distance transforms participate in the specification of the skeleton location.

1.2.9 Geometric Invariant Flows

The relation between the *Eulerian formulation* and curve evolution serves as a direct link between curve and image evolution. Under some limitations, like preserving the order of level sets (preserving the embedding), it is possible to evolve all the level sets simultaneously. Each of the evolving level sets will follow the same geometric evolution rule.

This important observation made it possible to extend the Euclidean scale space of planar curves [82, 74] into geometric image smoothing, the affine scale space of curves into affine invariant image smoothing [180, 4, 3], and the geodesic curvature flow into bending invariant smoothing.

1.2.10 Geodesic Active Contours

One of the main problems in image analysis is the segmentation problem. Given several objects in an image, it is necessary to integrate their boundaries in order to achieve a good model of the objects under inspection. This problem was addressed in many ways over the years, starting with simple thresholding, region growing, and deformable contours based on energy minimization along a given contour called "snakes" [204, 203].

In [34, 35] a novel geometric model that starts from a user-defined contour and segments objects in various type of images is introduced. The idea is to minimize a total "nonedge" penalty function $g(x, y)$ integrated along the curve. The relation to the classical snakes and to recent geometric models is explored, showing better behavior of the proposed method over its ancestors: the classical snakes [204, 203] and the recent geometric models [146, 145, 33]. The planar evolution of the boundary curve is

$$C_t = \left(\kappa g - \langle \nabla g, \vec{N} \rangle \right) \vec{N},$$

where κ is the curvature. It is the steepest descent flow derived from the EL of the weighted arclength functional

$$\int g(C)|C'|dp.$$

This way of finding local weighted geodesics in a potential function defined by an edge detection operator requires an initial contour as initial conditions. In some other cases, it is desired to locate the minimal geodesic connecting two points along the boundary of an object. In [47] an approach for integrating edges by an efficient search for the minimal geodesic is explored.

1.2.11   Minimal Surfaces and the Heat Equation

Plateau was the first to realize that a wire contour dipped into soapy water and glycerine forms minimal surfaces. More complicated situations appear as we dip two circular wires. The minimal surface formed between the two wires changes its topology from a cylinder connecting the two contours, which splits into two flat, circular regions as the distance between the contours is increased; see Figure 1.8. Questions about the mathematical properties, like existence, uniqueness, and stability, were addressed by many, and there are still many open questions. The question of how to simulate these effects numerically is also studied intensively. We will look into the results of Chopp and Sethian who applied the Osher–Sethian level set method to this problem. We will also apply techniques of surfaces that flow toward minimal ones for color image enhancement.

Figure 1.8: A soap film between two rings starting from a cylinder and splitting into two circular disks, which are the Plateau problem solutions for the circular wire contours.

Minimal surfaces are realized by a flow via the mean curvature of the surface in its normal direction. The "flows" we study involve an evolution equation that describes the infinitesimal change of some quantity, like a curve or a surface, in "time." When we apply this equation to a given initial curve, we obtain a modified one. We sometimes refer to the time as scale.

The simplest equation that results in *scale space* is the linear heat equation

$$u_t = u_{xx},$$

where $u(x; t = 0) = f(x)$ is given as initial data. This equation describes the heat profile in time along a metal rod; see Figure 1.9. In 2D it describes the effect of defocusing on an image in optics, as well as many other physical phenomena.

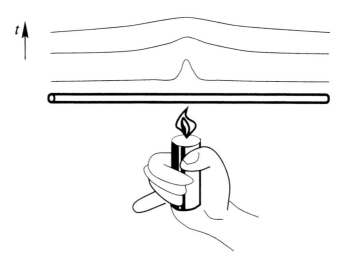

Figure 1.9: The heat profile along a metal rod in time.

The solution to the heat equation can be easily shown to be given by convolution of the initial data with a Gaussian kernel[1]

$$u(x;t) = g(x;t) * f(x) = \int_{-\infty}^{+\infty} g(\tilde{x};t)f(x - \tilde{x})d\tilde{x},$$

where

$$g(x;t) = \frac{1}{\sqrt{4\pi t}}e^{-x^2/(4t)}.$$

After the following exercises we will be ready to dive deeper into the world of evolving curves and surfaces, which we later apply to solve the various problems that we just reviewed.

1.3   Exercises

1. Compute the EL equations for the arclength functional of a parametric curve,

$$L(x'(p), y'(p)) = \int_0^1 \sqrt{x_p^2 + y_p^2}dp.$$

Compare with the EL equations of $L(x'(p), y'(p)) = \int_0^1 (x_p^2 + y_p^2)dp$.

---

[1]Prof. James A. Sethian from Berkeley gave Gaussian, Laplacian, Hamiltonian, Lagrangian, etc. as an example to the large number of Armenian mathematicians.

2. Compute the EL equations for the arclength functional of a closed parametric curve,

$$
\begin{aligned}
L(C(p), C'(p)) &= L(x(p), y(p), x'(p), y'(p)) \\
&= \int_0^1 g(x, y)\sqrt{x_p^2 + y_p^2}\, dp \\
&= \int_0^1 g(C)|C_p|\, dp.
\end{aligned}
$$

Express the EL with the geometric relations for the normal $\vec{N} = \{-y_p, x_p\}/\sqrt{x_p^2 + y_p^2}$ and the curvature $\kappa \equiv (x_p y_{pp} - y_p x_{pp})/(x_p^2 + y_p^2)^{3/2}$.

3. Compute the EL equations for the functional $\int F(u, u_x, u_{xx}, \dots)\, dx$.

4. Compute the EL equations for the functional $\iint F(u, u_x, u_y, u_{xy}, u_{xx}, u_{yy})\, dx\, dy$.

5. Show that given the function $F(y_1, y_2, \dots, y_n)$ that defines the functional

$$
E(u) = \int F(c_1(u), c_2(u), \dots, c_n(u))\, dx
$$

where c_i are linear operators, then the Euler–Lagrange equations are given by $\sum_i^n c_i^*(dF/dy_i)$. Here, c^* is the conjugate operator of c defined in the classical way, $\langle v, cu \rangle = \langle c^* v, u \rangle$.[2]

6. We saw that the *length* element of a function $y(x)$ is given by $ds = \sqrt{1 + y_x^2}\, dx$. Show that the area element of a function $z(x, y)$ is given by $da = \sqrt{1 + z_x^2 + z_y^2}\, dx\, dy$. Next, compute the EL equations for the functional $\int da$. (This is Joseph-Louis Lagrange's result from 1788.)

7. Prove that Heron's formula approximates the curvature radius.

   The curvature radius at a given point on a curve is the radius of the osculating circle at this point. Assume a curve is sampled by points. Prove that the radius of the circle defined by three points A, B, C is given by $r(B) = abc/(4\sqrt{s(s-a)(s-b)(s-c)})$, where a is the distance between A and B denoted by $a = d(B, C)$, $b = d(A, C)$, $c = d(A, B)$, and $s = (a + b + c)/2$. (This is Heron of Alexandria's result

---

[2] This nice and compact approach to calculus of variations was introduced by Avraham Levy of HP Labs, Israel.

from 90. It's actually believed to have been known to Archimedes before 212 B.C.)

Hint: Use the law of cosines to prove that the area is given by $\sqrt{s(s-a)(s-b)(s-c)}$.

8. Prove that $g(x;t) * f(x)$ with the Gaussian kernel solves the heat equation $u_t = u_{xx}$.

9. Prove the *maximum principle* for the heat equation. That is, prove that

$$\inf_x u(x;0) \le g(t) * u(x;0) \le \sup_x u(x;0).$$

Hint: Prove and then use the facts that $\int_{\mathbb{R}^N} g(\tilde{x};t)d\tilde{x} = 1$ and that

$$\forall x_0 > 0, \qquad \lim_{t \to 0} \int_{x \in \mathbb{R}^N, |\tilde{x}| > x_0} g(\tilde{x};t)d\tilde{x} = 0.$$

10. What is the kernel for the solution of the general linear heat equation for a signal in an arbitrary dimension $u(x_1, x_2, \ldots, x_N; t)$?

Formally, we say that $u : \mathbb{R}^N \times \mathbb{R}_+ \to \mathbb{R}$, where $\mathbb{R}^N$ is the N dimensional Euclidean space span by the x_i arguments, and $\mathbb{R}_+$ is the ray of positive values that describes the time or scale. The arbitrary dimension linear heat equation is

$$u_t = \Delta u,$$

where $\Delta u = u_{x_1 x_1} + u_{x_2 x_2} + \cdots + u_{x_N x_N}$ is the Laplacian.

11. What is the kernel of the 2D linear affine heat equation

$$u_t = \text{div}\,(M\nabla u),$$

where M is a positive definite, symmetric matrix

$$M = \begin{pmatrix} a & c \\ c & b \end{pmatrix}?$$

Plot the level sets (also known as iso-contours, or equal-height contours) of the kernel.

12. Apply the 2D linear heat equation $u_t = u_{xx} + u_{yy}$ to a gray-level image $u(x, y; 0) = I(x, y)$. Program either the explicit form with forward Euler numerical iterations in time and central derivative approximation in space, or by a direct multiplication in the frequency domain, by multiplying the transformed signal with the transformed Gaussian kernel. What is the transform of the Gaussian kernel?

For the explicit form, denote $U_{i,j}^n = u(idx, jdy; ndt)$; then the iterative update scheme is given formally by

$$D_+^t U_{i,j}^n \;\; = \;\; (D_+^x D_-^x + D_+^y D_-^y) U_{i,j}^n,$$

where the definitions of backward and forward derivatives will be given in Chapter 5, or explicitly

$$U_{i,j}^{n+1} \;\; = \;\; U_{i,j}^n + dt \left(\frac{U_{i+1,j}^n - 2U_{i,j}^n + U_{i-1,j}^n}{dx^2} \right. $$
$$\left. + \; \frac{U_{i,j+1}^n - 2U_{i,j}^n + U_{i,j-1}^n}{dy^2} \right).$$

13. Repeat Exercise 12 for $u_t = \operatorname{div}(g\nabla u)$, where $g = (1 + |\nabla u|^2/\lambda^2)^{-1}$. Try different values for $\lambda > 0$. Apply forward derivatives for ∇u, and backward for the div, where the forward derivative is defined as $D_+^x U_{i,j} \equiv (U_{i+1,j} - U_{i,j})/dx$, and the backward by $D_-^x U_{i,j} \equiv (U_{i,j} - U_{i-1,j})/dx$.

2

Basic Differential Geometry

In this chapter we learn some basics of differential geometry of planar curves and curved surfaces that we later use in our applications.

2.1 Introduction to Differential Geometry in the Plane

Let us start with a closed simple parametric curve in the plane. We say that a curve is simple if it does not intersect itself. Let our curve be given by $C(p) = \{x(p), y(p)\}$, where $p \in [0, 1]$. Formally, we say that the curve C maps the interval $I = [0, 1]$ to the Euclidean plane $\mathbb{R}^2$ and write $C : I \rightarrow \mathbb{R}^2$. Here, p is the parameter of the curve where for every value of p between 0 and 1, $C(p)$ is the coordinate of one point along the curve.

The first derivative of the curve with respect to (or for short w.r.t.) its parameter p is the *velocity* or *tangent vector*, which we denote as

$$T = \frac{\partial C(p)}{\partial p} = C_p = \{x_p, y_p\}.$$

The *acceleration* vector is the derivative of the velocity according to the parameter p, or the second derivative of the curve C_{pp}. The unit tangent is obtained by normalization such that

$$\vec{T} = \frac{T}{|T|} = \frac{\{x_p, y_p\}}{\sqrt{x_p^2 + y_p^2}}.$$

Given a vector in the plane, the normal to this vector is obtained by changing the positions of its arguments and inverting the sign of one of them.

For example, the unit normal to a curve is given by

$$\vec{N} = \frac{\{-y_p, x_p\}}{\sqrt{x_p^2 + y_p^2}};$$

see Figure 2.1. The inner product between the tangent and the normal is always zero (by the construction of the normal),

$$\langle T, \vec{N} \rangle = x_p \frac{-y_p}{|C_p|} + y_p \frac{x_p}{|C_p|} = 0.$$

We sometimes also use the dot sign for the inner product (e.g., $\vec{T} \cdot \vec{N} = 0$).

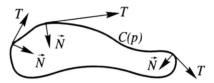

Figure 2.1: The tangent $T(p)$ and the normal $\vec{N}(p)$ along the parametric curve $C(p)$.

The length from 0 to p along a curve may be measured via

$$s(p) = \int_0^p |C_{\tilde{p}}(\tilde{p})| d\tilde{p} \quad = \quad \int_0^p \sqrt{x_{\tilde{p}}^2(\tilde{p}) + y_{\tilde{p}}^2(\tilde{p})} d\tilde{p},$$

and in differential form we have that $ds^2 = (x_p^2 + y_p^2)dp^2$, or $ds = |C_p|dp$, and we also state that $dp/ds = 1/|C_p|$. Note that the function s maps the parameterization interval of p given by $I = [0, 1]$ to a new interval given by $[0, L]$ where $L = \int_0^1 |C_p|dp$ is the total length of the curve. For $ds/dp = |C_p| > 0$, that is $s(p)$ is strictly increasing, we have that $s : [0, 1] \rightarrow [0, L]$ is one-to-one and thus has an inverse $p(s)$.

Now that we have an *intrinsic* distance measure along the curve, we can *change the parameterization* to the Euclidean arclength. That is, we set the velocity along the curve to be one. We say, by abuse of notation, that $C(s) = C(p(s))$ is a curve with arclength parameter. By applying the chain rule we have $C_s = C_p \partial_s p = C_p/|C_p|$, so that indeed we have a unit velocity $|C_s| = 1$.

The arclength parameterization and the curve derivatives according to it are intrinsic, geometric, and invariant to Euclidean transformation of the coordinates. We already saw that $C_s = \vec{T} = C_p/|C_p|$. The second derivative has also a very special meaning. Let us compute C_{ss}.

$$
\begin{aligned}
C_{ss} = \partial_s C_s = \frac{\partial}{\partial s}\frac{C_p}{|C_p|} &= \left(\frac{\partial}{\partial p}\frac{C_p}{|C_p|}\right)\frac{dp}{ds} \\
&= \left(\frac{C_{pp}}{|C_p|} - \frac{C_p(x_p x_{pp} + y_p y_{pp})}{|C_p|^3}\right)\frac{1}{|C_p|} \\
&= \frac{C_{pp}|C_p|^2 - C_p(x_p x_{pp} + y_p y_{pp})}{|C_p|^4} \\
&= \frac{x_p y_{pp} - y_p x_{pp}}{(x_p^2 + y_p^2)^{3/2}}\frac{\{-y_p, x_p\}}{|C_p|} \\
&= \frac{(C_p, C_{pp})}{|C_p|^3}\vec{N}.
\end{aligned}
$$

We introduced new notation

$$
(\{a,b\},\{c,d\}) = \det\begin{pmatrix} a & b \\ c & d \end{pmatrix} = ad - bc.
$$

For two vectors in the plane, it is the outer product that is equal to the area (up to a sign) of the parallelogram defined by the two vectors as edges. The quantity $\kappa = (C_p, C_{pp})/|C_p|^3$ is called the curvature, so we have $C_{ss} = \kappa\vec{N}$. Note that by differentiating $\langle C_s, C_s\rangle = 1$ with respect to s, we have $\langle C_{ss}, C_s\rangle = 0$, which means that the first and second derivatives of the curve are orthogonal. For example, for $C(x) = \{x, y(x)\}$ we obtain $\kappa = -y_{xx}/(1 + y_x^2)^{3/2}$, which we already saw in Chapter 1. The curvature $\kappa = 1/\rho$, where ρ is called the curvature radius, which is the radius of the osculating circle. It is the circle uniquely defined by taking three neighboring points along the curve and letting the arclength between the points go to zero; see Figure 2.2.

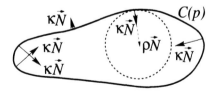

Figure 2.2: The curvature $\kappa = 1/\rho$, where $C_{ss} = \kappa\vec{N}$.

The curvature magnitude is also defined as the derivative w.r.t. s of the angle between the tangent and the $\hat{x} = \{1, 0\}$ axis, $\theta \equiv \angle(\vec{T}, \hat{x})$. Let us follow the notations in Figure 2.3 for the following geometric proof. As $ds \to 0$, we have that $\sin d\theta = d\theta = ds/\rho$ such that $d\theta/ds = 1/\rho$, and thus $\kappa = \theta_s$. Note that at the same token we proved that $\vec{T}_s = \kappa\vec{N}$: Set the complex notation $\vec{T}(s) = e^{i\theta(s)} \equiv \cos(\theta(s)) + i\sin(\theta(s))$. Then, $d\vec{T}/ds = i\theta_s e^{i\theta(s)} = \kappa i e^{i\theta(s)} = \kappa\vec{N}$.

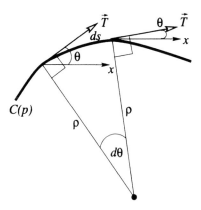

Figure 2.3: The curvature is defined as $\kappa = \theta_s$.

2.2   Invariant Signatures

In this section we explore *invariant differential signatures*, which are useful for pattern recognition and object classification under partial occlusion. We address the recognition problem based on local description of the object boundary; see, for example, [7, 8, 27, 28, 29, 48, 152, 180]. The goal is to identify a planar shape by differential analysis of its smooth boundary curve. We would like to learn the local properties of the boundary and extract a signature that will identify the shape we inspect under specific families of deformations. See Figure 2.4.

Figure 2.4: We try to fit the missing piece to the given hole boundary C^o. The question we would like to answer is how to represent the boundaries C^a, C^b, C^c, C^d that will enable us to compare the candidate pieces to the shape of the hole C^o.

2.2.1 Euclidean Invariants of Planar Curves

Euclidean invariant operations refer to those operations for which translations and rotations of objects in the plane do not affect the result of the operation. The simplest differential signature for the Euclidean group of transformations is the function $\kappa(s)$. It is equivalently given by the curve defined by the graph $\{s, \kappa(s)\}$. The Euclidean arclength and curvature are invariant to translations and rotations. Integrating $\kappa(s)$ once produces $\theta(s) + \theta_0$, where $\theta(s)$ is the angle between the tangent vector and the $\hat{x} = \{1, 0\}$ axis, and θ_0 is the arbitrary integration constant—which means reconstruction up to rotation. Integration of $\theta(s) = \arctan(y_s/x_s)$ results in $\{x(s), y(s)\} + \{x_0, y_0\}$, where $\{x_0, y_0\}$ is the arbitrary integration constant, which means reconstruction up to translation. Practically, we solve the two coupled equations with two unknowns, dx and dy, for each s:

$$\begin{cases} ds^2 & = & dx^2 + dy^2, \\ \tan\theta(s) & = & \frac{dy}{dx}. \end{cases}$$

We next integrate for x and y; for example, $x(s) = \int_0^s x'(\tilde{s})d\tilde{s}$.

The curvature is defined as

$$\kappa = \frac{d\theta}{ds} = \frac{\frac{d\theta}{dp}}{\frac{ds}{dp}} = \frac{\theta_p}{\sqrt{x_p^2 + y_p^2}},$$

and by definition we also have that

$$\tan\theta = \frac{dy}{dx} = \frac{y_p}{x_p},$$

such that

$$\frac{d}{dp}(\tan\theta) = \sec^2\theta \frac{d\theta}{dp}$$

and

$$\begin{aligned} \theta_p & = & \cos^2\theta \frac{d}{dp}(\tan\theta) \\ & = & \left(\frac{dx}{ds}\right)^2 \frac{d}{dp}\left(\frac{y_p}{x_p}\right) \\ & = & (x_p p_s)^2 \frac{x_p y_{pp} - y_p x_{pp}}{x_p^2} \\ & = & \frac{x_p y_{pp} - y_p x_{pp}}{x_p^2 + y_p^2}. \end{aligned}$$

So we conclude with the curvature equation for an arbitrary parameterization

$$\kappa = \theta_s = \theta_p p_s = \frac{x_p y_{pp} - y_p x_{pp}}{x_p^2 + y_p^2}(x_p^2 + y_p^2)^{-1/2} = \frac{x_p y_{pp} - y_p x_{pp}}{(x_p^2 + y_p^2)^{3/2}}.$$

The signature $\kappa(s)$ describes the curve up to the starting point (shift) along the arclength parameterization. According to the Cartan theorem [157, 32], the signature $\{\kappa_s(s), \kappa(s)\}$ is a unique signature (without shifting) for any given planar curve. There is, however, practical numerical problems in the computation of more than a second-order derivative of a curve. For quantities that involve higher-order derivatives, the approximations require an extended numerical support that makes the measured quantity either nonlocal or very noisy. We will thus try to limit our applications to second-order derivatives.

2.2.2   Affine Invariants of Planar Curves

Let us start with basic concepts from the theory of affine differential geometry of planar curves. The affine group is the group of linear transformations given by the six-parameter transformation

$$\{\tilde{x}, \tilde{y}\}^T = A\{x, y\}^T + \mathbf{b} = \begin{pmatrix} a & b \\ c & d \end{pmatrix} \{x, y\}^T + \{b_1, b_2\}^T,$$

where $\det(A) > 0$. The equi-affine, or area-preserving affine group, adds the requirement that $\det(A) = 1$, which reduces the number of parameters to five. The interested reader is referred to [31] for more details on affine differential geometry.

We remind the reader that the inner and outer products are denoted as $\langle \{a, b\}, \{c, d\} \rangle \equiv ac + bd$, and $(\{a, b\}, \{c, d\}) \equiv ad - bc$, respectively. Let $C(p) : [0, 1] \rightarrow \mathbb{R}^2$ be a simple regular parametric planar curve: $C(p) = \{x(p), y(p)\}$. Let s be the Euclidean arclength so that

$$s(p) = \int_0^p \langle C_{\tilde{p}}, C_{\tilde{p}} \rangle^{1/2} d\tilde{p},$$

where $\tilde{p} \in [0, 1]$ is an arbitrary parameterization. The unit tangent is given by $\vec{T} = C_s$, and the Euclidean curvature vector is given by $C_{ss} = \kappa \vec{N}$, where $\vec{N}$ is the curve normal and κ is the Euclidean curvature. The equi-affine group of transformations is a subgroup of the affine in the sense of restricting the affine to preserve area. Given the general affine transformation $\hat{C} = AC + \mathbf{b}$, the restriction is $\det(A) = 1$.

The equi-affine arclength v is defined so that

$$(C_v, C_{vv}) = 1 \tag{2.1}$$

and is given by [31]:

$$v(p) = \int_0^p |(C_{\tilde{p}}, C_{\tilde{p}\tilde{p}})|^{1/3} d\tilde{p}; \tag{2.2}$$

see Figure 2.5, which is an *intrinsic integral* as we will see later.

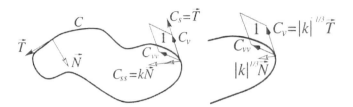

Figure 2.5: The geometric relations between C_v, C_{vv}, C_s, and C_{ss}, for v the affine arclength and s the Euclidean one.

Using the reparameterization invariance property of the integral in (2.2), the relation between s the Euclidean and v the affine arclength [181] is obtained from

$$v = \int |(C_s, C_{ss})|^{1/3} ds,$$

which yields

$$\begin{aligned}
\frac{dv}{ds} &= |(C_s, C_{ss})|^{1/3} \\
&= |(\vec{T}, \kappa \vec{N})|^{1/3} \\
&= |\kappa|^{1/3}.
\end{aligned} \tag{2.3}$$

Using this expression, the *affine tangent* is given by

$$C_v = C_s \frac{\partial s}{\partial v} = |\kappa|^{-1/3} \vec{T}.$$

Differentiating Eq. (2.1), we have

$$(C_v, C_{vvv}) = 0, \tag{2.4}$$

which means that the vectors C_v and C_{vvv} are linearly dependent: $C_{vvv} = -\mu C_v$. The scalar μ is the simplest affine differential invariant of the curve C, known as the *affine curvature*, and C_{vv} is the affine normal vector. A direct result from the last equation is

$$\mu = (C_{vv}, C_{vvv}).$$

Differentiating Eq. (2.4) with respect to v, it also follows that

$$\mu = (C_{vvvv}, C_v).$$

Observe that

$$
\begin{aligned}
C_{vv} &= \frac{\partial}{\partial v}\left(C_s\frac{\partial s}{\partial v}\right) \\
&= C_{ss}\left(\frac{\partial s}{\partial v}\right)^2 + C_s\frac{\partial^2 s}{\partial v^2} \\
&= \kappa\vec{N}|\kappa^{-1/3}|^2 + \vec{T}\frac{\partial^2 s}{\partial v^2} \\
&= \kappa^{1/3}\vec{N} + \vec{T}\frac{\partial}{\partial v}|\kappa^{-1/3}| \\
&= \kappa^{1/3}\vec{N} - \frac{\kappa_s}{3|\kappa|^{5/3}}\vec{T},
\end{aligned}
$$

a relation that will be found useful in the forthcoming chapter.

Let $g : \mathbb{R}^2 \to \mathbb{R}^+$ be a positive weight function defined on the plane. Without breaking the affine invariance property, we integrate g along the affine arclength. Let

$$
L(C, C_p, C_{pp}) = g(C(p))(C_p, C_{pp})^{1/3}.
$$

We denote the following functional as the "weighted affine arclength":

$$
\mathcal{L}(C) = \int L(C, C_p, C_{pp})dp. \tag{2.5}
$$

Let us prove that (2.5) is free of parameterization [75] (i.e., an intrinsic integral), a property that should hold for any arclength.

Lemma 1 *The functional*

$$
\int_{p_0}^{p_1} g(C(p))(C_p, C_{pp})^{1/3}dp
$$

depends only on the curve in the xy-plane, also referred to as the "orbit," "trace," or "image" of the curve $C(p)$, defined by the parametric equation $x = x(p)$, $y = y(p)$, and not on the choice of the parametric representation of the curve.

Proof. We show that if we go from p to a new parameter r by setting $p = p(r)$, where $dp/dr > 0$ and the interval $[p_0, p_1]$ goes into $[r_0, r_1]$, then

$$
\int_{r_0}^{r_1} g(C(r))(C_r, C_{rr})^{1/3}dr = \int_{p_0}^{p_1} g(C(p))(C_p, C_{pp})^{1/3}dp.
$$

Since we have $y_r = y_p p_r$, $y_{rr} = y_{pp}p_r^2 + y_p p_{rr}$, and similarly $x_r = x_p p_r$, $x_{rr} = x_{pp}p_r^2 + x_p p_{rr}$, it follows that

$$\int_{r_0}^{r_1} g(C(r))(C_r, C_{rr})^{1/3} dr$$

$$= \int_{r_0}^{r_1} g(C(r))\big(x_p p_r (y_{pp} p_r^2 + y_p p_{rr})$$

$$- y_p p_r (x_{pp} p_r^2 + x_p p_{rr})\big)^{1/3} dr$$

$$= \int_{r_0}^{r_1} g(C(r)) \left(x_p y_{pp} p_r^3 - y_p x_{pp} p_r^3\right)^{1/3} dr$$

$$= \int_{r_0}^{r_1} g(C(r)) \left(x_p y_{pp} - y_p x_{pp}\right)^{1/3} \frac{dp}{dr} dr$$

$$= \int_{p_0}^{p_1} g(C(p))(C_p, C_{pp})^{1/3} dp. \qquad \blacksquare$$

Lemma 1 guarantees that the functional (2.5) is free of the parameterization of the curve. In fact, in the general case of selecting an arclength for a given transformation group, the integral must be intrinsic. It should have the general form of $dl = [\text{geometric quantity}] \times |C_p| dp$, where "geometric quantity" is 1 for the Euclidean group, $|\kappa(p)|^{1/3}$ for the equi-affine group, $|\kappa(p)|$ for the similarity, $|(C(p), \vec{T}(p))|$ for the linear-affine, etc.

The relation between the affine curvature μ and the curvature derivatives κ, κ_s, and κ_{ss} is obtained by minimizing the affine arclength and using the resulting Euler–Lagrange equations that yield

$$\mu = \kappa^{4/3} - \frac{5}{9}\kappa^{-8/3}\kappa_s^2 + \frac{1}{3}\kappa^{-5/3}\kappa_{ss}. \qquad (2.6)$$

This equation can be found in other ways (see, e.g., [31, 152, 174]). Note that the affine curvature μ involves fourth-order derivatives, which makes it a numerically sensitive signature measure.

The following table presents invariant arclengths and their corresponding geometric heat equations (which we explore in Chapter 3) for some simple transformation groups.

Group Arclength	$L(p)$	Geom. Heat Eq.
Euclidean	$\langle C_p, C_p \rangle^{1/2} = 1 \cdot |C_p|$	$C_t = \kappa \vec{N}$
Weighted Euclidean	$g(C)\langle C_p, C_p \rangle^{1/2}$	$C_t = \frac{\kappa}{g^2}\vec{N}$
Equi-affine	$|(C_p, C_{pp})|^{1/3} = |\kappa|^{1/3}|C_p|$	$C_t = \kappa^{1/3}\vec{N}$
Weighted affine	$g(C)|(C_p, C_{pp})|^{1/3} = g(C)|\kappa|^{1/3}|C_p|$	$C_t = \frac{1}{g^2}\kappa^{1/3}\vec{N}$
Similarity	$\frac{|(C_p, C_{pp})|}{\langle C_p, C_p \rangle} = |\kappa||C_p|$	$C_t = \frac{1}{\kappa}\vec{N}$
Linear-equi-affine	$|(C, C_p)| = |(C, \vec{T})||C_p|$	$C_t = \frac{\kappa}{(C, \vec{T})^2}\vec{N}$
Weighted linear-equi-affine	$g(C)|(C, C_p)| = g(C)|(C, \vec{T})||C_p|$	$C_t = \frac{\kappa}{g^2(C, \vec{T})^2}\vec{N}$
Linear affine	$\frac{|(C_p, C_{pp})|}{(C, C_p)^2} = \frac{|\kappa|}{(C, \vec{T})^2}|C_p|$	$C_t = \frac{(C, \vec{T})^4}{\kappa}\vec{N}$
	$\frac{|(C, C_{pp})|}{|(C, C_p)|} = \left|\frac{\kappa(C, \vec{N})}{(C, \vec{T})}\right||C_p|$	$C_t = \frac{(C, \vec{T})^2}{\kappa(C, \vec{N})^2}\vec{N}$

2.3   Calculus of Variations in Parametric Form

The measures we care about are those that do not depend on a specific parameterization of the problem at hand. Let us follow [2] and introduce simple requirements for such geometric functionals like the Euclidean arclength. For the case of a parametric curve, we require that the functional

$$J(C) = \int_C F(p, C(p), C_p(p)) dp = \int_{p_1}^{p_2} F(p, C(p), C_p(p)) dp$$

depends only on the curve C, and not on the parameterization. By this "geometric" property, we actually require that for any monotonic reparameterization $q = w(p)$, we have

$$
\begin{aligned}
\int_{p_1}^{p_2} F(p, C, C_p) dp &= \int_{q_1}^{q_2} F(q, \tilde{C}(q), \tilde{C}_q(q)) dq \\
&= \int_{p_1}^{p_2} F\left(w(p), C(w(p)), \frac{C_p(w(p))}{w_p(p)}\right) w_p(p) dp.
\end{aligned}
$$

Thus,

$$F(p, C, C_p) = F\left(w(p), C(w(p)), \frac{C_p(w(p))}{w_p(p)}\right) w_p(p)$$

is a necessary and sufficient condition for our "reparameterization invariance".

Let us verify the condition for $w(p) = p + c$, for some constant c. This choice of reparameterization yields the following restriction:

$$F(p, C, C_p) = F(C, C_p).$$

Now, for $w(p) = cp$ where c is a positive constant, we obtain

$$F(C, cC_p) = cF(C, C_p).$$

In the remainder of this section we will use the short-hand notation $C_p = \{x', y'\}$. If we differentiate both sides with respect to c and set $c = 1$, we get

$$F_{x'} x' + F_{y'} y' = F. \tag{2.7}$$

Differentiating once w.r.t. x', and once w.r.t. y', we get

$$
\begin{aligned}
x' F_{x'x'} + y' F_{y'x'} &= 0, \\
x' F_{x'y'} + y' F_{y'y'} &= 0.
\end{aligned}
$$

We conclude that

$$F_1 \equiv \frac{F_{x'x'}}{y'^2} = -\frac{F_{x'y'}}{x'y'} = \frac{F_{y'y'}}{x'^2}. \tag{2.8}$$

The curve $C(p)$ should satisfy the Euler–Lagrange equations

(a) $\dfrac{d}{dp}F_{x'} - F_x = 0,$

(b) $\dfrac{d}{dp}F_{y'} - F_y = 0.$

Equation (a) can be written as

$$F_{x'x}x' + F_{x'y}y' + F_{x'x'}x'' + F_{x'y'}y'' = F_x. \tag{2.9}$$

Differentiating Eq. (2.7) w.r.t. x, we get

$$F_x = F_{xx'}x' + F_{xy'}y'.$$

Now, using the definition (2.8), we can write (2.9) as

$$y'\left((F_{x'y} - F_{xy'}) + F_1(x''y' - y''x')\right) = 0.$$

Applying a similar procedure, (b) can be written as

$$x'\left((F_{x'y} - F_{xy'}) + F_1(x''y' - y''x')\right) = 0.$$

Since x' and y' cannot vanish simultaneously, we have the result

$$F_{x'y} - F_{xy'} + F_1(x''y' - y''x') = 0,$$

known as the *Euler–Weierstrass equation*. For $F_1 \neq 0$ it can be written in the following geometric form:

$$\frac{F_{x'y} - F_{xy'}}{F_1(x'^2 + y'^2)^{3/2}} = \frac{x'y'' - y'x''}{(x'^2 + y'^2)^{3/2}} \equiv \kappa,$$

where κ is the well-known curvature.

2.4 Geometry of Surfaces

In the last sections we studied the properties of curves in the plane. This section will take us to higher dimensions. Curves are not restricted to live in a plane. Although one-dimensional "creatures," curves can be "embedded" in any Euclidean space, which we also call "flat manifold," or in any nonflat manifold. It is easy to think of a wire in 3D.

Two-dimensional surfaces can also be embedded in spaces with arbitrary dimensions, though it is harder to imagine a two-dimensional surface hovering in a four-dimensional space. For example, a 2D parameterized surface embedded in 4D flat space,

$$\mathcal{S}(u,v) = \{w(u,v), x(u,v), y(u,v), z(u,v)\},$$

is a mapping of the (u, v) parameterization plane to the 4D space $\mathcal{S} : \mathbb{R}^2 \to \mathbb{R}^4$.

Before we go into applications let us introduce some useful definitions. A nice dictionary for differential geometry and general mathematical vocabulary is [214], from which we adopted some of the following explanations.

- Manifold

  Any object that can be locally charted is called a manifold. More formally, a manifold M is a topological space for which any point has a neighborhood $U \subset M$ that can be mapped into a Euclidean space. This mapping should be continuous, one-to-one, onto and have a continuous inverse. A mapping with such properties is referred to as a homeomorphic mapping, where the homeomorphism is called a chart. In other words, the neighborhood U should be homeomorphic to a Euclidean space of the same dimension as that of the manifold M. If we restrict ourselves to two-dimensional manifolds, also called surfaces, then the chart is a map to a plane, like charts or maps of small regions of the Earth.

  We will denote Euclidean spaces of dimension n as $\mathbb{R}^n$. A smooth manifold of dimension n with a smooth parameterization is a differentiable manifold for which overlapping charts relate smoothly. This means that the inverse of one chart followed by another is an infinitely differentiable map from $\mathbb{R}^n$ to itself.

- Riemannian manifold

  A manifold with a metric tensor is called a Riemannian manifold. The metric tensor will be defined in the next section. A Riemannian manifold is complete if the distance $d(q, r)$ between two manifold points q and r is defined as the length of the minimal geodesic (shortest curve in the manifold) connecting q to r.

- Distance

  Let $C(p)$ be a smooth curve in a manifold M connecting q to r with $C(0) = q$ and $C(1) = r$. Then, $C'(p) \in T_{C(p)}$, where T_q is the tangent hyperplane of M at q. The length of C with respect to the Riemannian structure is given by

  $$\int_0^1 |C'(p)| dp,$$

  where $|C'| = \langle C', C' \rangle_g^{1/2}$ will be explicitly defined in the following section. The distance $d(q, r)$ between q and r is the shortest distance between q and r given by

$$d(q, r) = \inf_{C\,:\,\text{from } q \text{ to } r} \int |C'(p)| dp;$$

$d(q, r)$ is called the "geodesic distance" when M is not flat.

- Tangent plane

  A tangent plane T_p is a plane that meets a given surface at a single point.
  Example: For a two-dimensional parameterized surface $S(u, v)$ embedded in $\mathbb{R}^3$, the tangent plane at each point is defined by the two tangents S_u and S_v; see Figure 2.6.

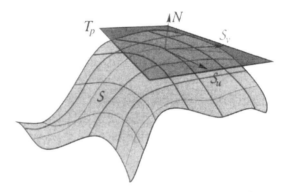

Figure 2.6: T_p, the tangent plane, contains the two tangents S_u and S_v.

  Example: Let (x_0, y_0) be a point of a surface function $\{x, y, z = f(x, y)\}$. The tangent plane T_p at (x_0, y_0) is given by x, y, z that satisfy

  $$z = f(x_0, y_0) + f_x(x_0, y_0)(x - x_0) + f_y(x_0, y_0)(y - y_0). \qquad (2.10)$$

- Geodesic curvature

  For an arclength parameterized curve $C(s)$ on a surface, the magnitude of the surface-tangential component of the acceleration C_{ss} is the geodesic curvature denoted by κ_g. Curves with $\kappa_g = 0$ are called *geodesics*.

  For example, let $S(u, v) = \{x(u, v), y(u, v), z(u, v)\}$ be a parameterized surface in $\mathbb{R}^3$, and consider the arclength parameterized curve

$C(s)$ embedded in $\mathcal{S}$. Then the geodesic curvature vector is defined as

$$\kappa_g \mathcal{N} = C_{ss} - \langle C_{ss}, \vec{N} \rangle \vec{N},$$

where $\vec{N}$ is the surface normal given by

$$\vec{N} = \frac{\mathcal{S}_v \times \mathcal{S}_u}{|\mathcal{S}_v \times \mathcal{S}_u|},$$

and the surface normal component of the acceleration C_{ss} is the normal curvature given by $\kappa_n = \langle C_{ss}, \vec{N} \rangle \vec{N}$; see Figure 2.7.

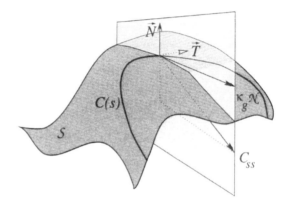

Figure 2.7: The geodesic curvature vector of a curve on a surface.

- Principal curvatures

  The maximum and minimum of the normal curvature κ_1 and κ_2 at a given point on a surface are called the principal curvatures. They measure the maximum and minimum bending of the surface at each point. The *Gaussian curvature K* and *mean curvature H* are defined by

$$
\begin{aligned}
K &= \kappa_1 \kappa_2, \\
H &= \frac{1}{2}(\kappa_1 + \kappa_2).
\end{aligned}
$$

  The relation between K, H, κ_1, and κ_2 is captured by the quadratic equation $\kappa^2 - 2H\kappa + K = 0$, with the solutions

$$
\begin{aligned}
\kappa_1 &= H + \sqrt{H^2 - K}, \\
\kappa_2 &= H - \sqrt{H^2 - K}.
\end{aligned}
$$

- Surface area

  The area of a parameterized surface $\mathcal{S}(u, v)$ is an intrinsic quantity, defined by

  $$\mathcal{A} = \int_{\mathcal{S}} |\mathcal{S}_u \times \mathcal{S}_v| du dv,$$

  where $\mathcal{S}_u$ and $\mathcal{S}_v$ are the tangent vectors, and $\mathcal{S}_u \times \mathcal{S}_v$ is their cross product. The magnitude of the cross product of two vectors is the area of the parallelogram for which the vectors are edges. See Figure 2.8.

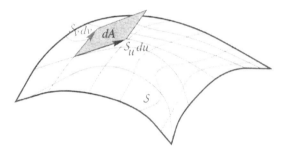

Figure 2.8: A surface area element dA is defined by the cross product of the tangent vectors generated w.r.t. the parameterization.

2.5 A Brief Introduction to Intrinsic Geometry

Let us define again the very basic measure, which is the arclength in terms of its metric. The most general form is given by

$$ds^2 = g_{ij} dX^i dX^j,$$

where X^i are the elements of a parametric representation of our embedded manifold, and here we use Einstein summation convention: Identical indices that appear one up and one down are summed over. For example, in the Euclidean plane $\mathbb{R}^2$, where $X^1 = x$, and $X^2 = y$, we can compute the metric coefficients from the classical Euclidean arclength definition,

$$ds^2 = dx^2 + dy^2 = \delta_{ij} dX^i dX^j.$$

Here

$$g_{ij} = \delta_{ij} \equiv \left\{ \begin{array}{ll} 1 & i = j, \\ 0 & i \neq j. \end{array} \right.$$

As we go to surfaces, we need to apply a similar procedure. For example, consider the parameterized surface $\mathcal{S}(x,y) = \{x, y, z(x,y)\}$, embedded in the Euclidean space $\mathbb{R}^3$. The manifold can be written by the elements of its parametric representation $\mathcal{S} = \{X^1, X^2, X^3\}$, where in our example $X^1(x,y) = x$, $X^2(x,y) = y$, and $X^3(x,y) = z(x,y)$. In this case, the arclength on the surface is given by $ds^2 = dx^2 + dy^2 + dz^2$. Applying the chain rule, we have

$$dz(x,y) = \frac{dz}{dx}dx + \frac{dz}{dy}dy = z_x dx + z_y dy.$$

Plugging $dz(x,y)$ into the arclength definition, we have

$$\begin{aligned}
ds^2 &= dx^2 + dy^2 + dz^2 \\
&= dx^2 + dy^2 + (z_x dx + z_y dy)^2 \\
&= (1 + z_x^2)dx^2 + 2z_x z_y dx dy + (1 + z_y^2)dy^2,
\end{aligned}$$

from which we can extract the metric coefficients

$$(g_{ij}) = \begin{pmatrix} g_{11} & g_{12} \\ g_{21} & g_{22} \end{pmatrix} = \begin{pmatrix} 1 + z_x^2 & z_x z_y \\ z_x z_y & 1 + z_y^2 \end{pmatrix}.$$

Again, the arclength can be written as $ds^2 = g_{ij}dX^i dX^j$, where by the parameterization we have chosen, $X^1 = x$, and $X^2 = y$. See Figure 2.9.

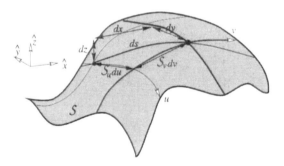

Figure 2.9: The squared arclength on a surface is given by a positive definite symmetric bilinear form, $ds^2 = g_{11}du^2 + 2g_{12}dudv + g_{22}dv^2 = dx^2 + dy^2 + dz^2$, called the metric (g_{ij}).

The *inverse metric*, denoted as $(g^{ij}) \equiv (g_{ij})^{-1}$, for the above example is given by

$$(g^{ij}) = \frac{1}{g}\begin{pmatrix} 1 + z_y^2 & -z_x z_y \\ -z_x z_y & 1 + z_x^2 \end{pmatrix},$$

where $g = \det(g_{ij}) = g_{11}g_{22} - g_{12}^2$.

In the more general case of a parameterized surface still in $\mathbb{R}^3$, given by $\mathcal{S} = \{x(u,v), y(u,v), z(u,v)\}$, we can use the arclength definition to write the area element in a compact way as

$$dA = \sqrt{g}\,dudv.$$

For simplicity, let us restrict ourselves to surfaces embedded in Euclidean spaces, in which case the metric is given by

$$(g_{ij}) = (\delta_{kl}\partial_i X^k \partial_j X^l) = \begin{pmatrix} x_u^2 + y_u^2 + z_u^2 & x_u x_v + y_u y_v + z_u z_v \\ x_u x_v + y_u y_v + z_u z_v & x_v^2 + y_v^2 + z_v^2 \end{pmatrix}.$$

Recall that the area element is defined as

$$dA \equiv \left| \frac{\partial \mathcal{S}}{\partial u} \times \frac{\partial \mathcal{S}}{\partial v} \right| dudv,$$

from which it is straightforward to show that $dA = \sqrt{\det(\partial_i X^k \partial_j X^k)}\,dudv$.

So, after we define the arclength ds and extract the appropriate metric (g_{ij}), we readily have an area measure given by the Nambu functional

$$A = \int \sqrt{g}\,dudv.$$

If we vary this functional according to X^i, we obtain a set of EL equations that forces the mean curvature vector to vanish everywhere,

$$\mathbf{H} \equiv H\vec{N} = 0.$$

The mean curvature vector is given by

$$\mathbf{H} \equiv \frac{1}{\sqrt{g}}\partial_i(\sqrt{g}g^{ij}\partial_j \mathcal{S}) = \Delta_g \mathcal{S},$$

where Δ_g is called the *second differential operator of Beltrami* or *Laplace–Beltrami operator*. It is an extension of the usual Laplacian operator to curved manifolds, where g stands for the metric on the manifold.

We proved in the first chapter that the first variation of the area of a surface yields the Euler–Lagrange equations $H\vec{N} = 0$, which means that a surface of minimal area should have zero mean curvature. Examples for such surfaces are planes, where the principal curvatures are both zero, or any regular surface where every point is a hyperbolic point with negative Gaussian curvature such that $\kappa_1 = -\kappa_2$.

2.6 Exercises

1. Prove the invariance property of the Euclidean arclength and the curvature. Prove that the general Euclidean transformation (that

includes rotation and translation) applied to a curve $C(p) = \{x(p), y(p)\}$ does not change its arclength and curvature.

Hint: Show that the curvature along the curve $\tilde{C}(p) = \mathbf{R}\,C^T(p) + \mathbf{b}$ is the same as that of $C(p)$, where $\mathbf{R} = \begin{pmatrix} \cos\alpha & -\sin\alpha \\ \sin\alpha & \cos\alpha \end{pmatrix}$ is a rotation matrix, and $\mathbf{b} = \{b1, b2\}^T$ is a constant translation vector.

2. Show that the affine arclength equation $v(p) = \int_0^p (C_{\tilde{p}}, C_{\tilde{p}\tilde{p}})^{1/3} d\tilde{p}$ can be extracted from the fixed-area demand $(C_v, C_{vv}) = 1$.

3. Find the simplest invariant arclength to the shear transformation $\tilde{C}(p) = \mathbf{S}\,C^T(p) + \mathbf{b}$, where $\mathbf{S} = \begin{pmatrix} 1 & a \\ 0 & 1 \end{pmatrix}$, and $\mathbf{b} = \{b_1, b_2\}^T$ is a constant translation vector. What are the tangent, normal, and curvature in this case? Compute the invariant signature for a nonconvex points sampled curve. Report numerical difficulties. Verify experimentally that your signature is indeed invariant to the shear transformation.

4. Use the Euler–Weierstrass equation to extract the curvature of the "shortest path" for the weighted arclength case $L = \int w(C(p))|C_p| dp$.

5. Compute the EL equation for the area measure of a 2D surface embedded in $\mathbb{R}^5$, and parameterized by the x, y-coordinates $\{x, y, r(x, y), g(x, y), b(x, y)\}$.

6. Basic calculus: Prove the tangent plane Eq. (2.10) for the graph surface $\{x, y, f(x, y)\}$.

7. Basic calculus: Prove Green's theorem. Use the theorem to find a different formula for the area of a shape bounded by the curve $C = \{x, y(x)\}$. Which formula is more appropriate for the area computation of a planar polygon?
   Reminder: Green's theorem states that $\iint_\Omega (g_x - f_y) dx dy = \oint_{\partial\Omega = C} (f dx + g dy)$.

8. A small optional research project: Apply all versions of invariant differential signatures to identify smooth shapes according to their boundaries.

3

Curve and Surface Evolution

3.1 Evolution

At this point we have enough vocabulary to introduce motion to our curves. We add a new variable, t, to the curve and assume the curve is the first in a family of curves $C(p,t)$. One compact way of describing this family is by a differential evolution rule. For example,

$$\partial_t C(p,t) = \vec{N}$$

is a differential evolution rule that describes a curve moving along its normal direction with a constant velocity. The family of curves obtained this way are referred to as offsets; see Figure 3.1.

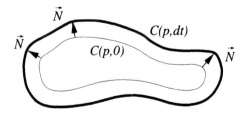

Figure 3.1: A curve moving along its normal with constant velocity.

As a second example let us consider the curvature flow, also known as the "geometric heat equation",

$$\partial_t C(p,t) = \kappa \vec{N}.$$

Again, we have a differential evolution rule that describes a curve moving along its normal direction with a velocity equal to its curvature. The reason for the name "geometric heat equation" comes from rewriting the curvature flow as $C_t = C_{ss}$ for s the Euclidean arclength. The family of curves obtained this way has nice properties; see Figure 3.2. Before we focus on

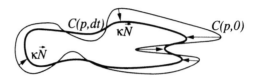

Figure 3.2: A curve moving along its normal with $C_{ss} = \kappa \vec{N}$.

the properties of the curvature flow let us introduce one important result for general geometric curve evolution. Consider the family of planar curves given by $C(s,t) : [0, \mathcal{L}(t)] \times [0, T) \to \mathbb{R}^2$, where s is the arclength of the curve C at time t, and $\mathcal{L}(t)$ is the length of the curve at time t. Let the curve evolution equation describing the differential change of the curve in time be given by

$$C_t = \vec{V}, \qquad C(s,0) = C_0(s),$$

where $\vec{V}(s,t) : [0, L] \times [0, T) \to \mathbb{R}^2$ is some velocity vector field that changes smoothly along the curve. The same evolution may be equivalently written by considering the normal $\vec{N} = \{-y_s, x_s\}$ and tangential $\vec{T} = C_s = \{x_s, y_s\}$ components of the velocity $\vec{V}$ along the curve:

$$C_t = \langle \vec{V}, \vec{N} \rangle \vec{N} + \langle \vec{V}, \vec{T} \rangle \vec{T}, \qquad C(s,0) = C_0(s).$$

See Figure 3.3.

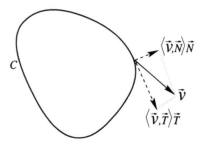

Figure 3.3: A velocity vector written in terms of its normal and tangential components.

A basic result from the theory of curve evolution states that the geometric shape of the curve, often referred to as the trace or the image of the planar curve, is only affected by the normal component of the velocity. The tangential component affects only the parameterization, and not the geometric shape of the propagating curve:

Lemma 2 [Epstein–Gage] [67]: *The family of curves $C(p,t)$ that solve the evolution rule*

$$C_t = V_N \vec{N} + V_T \vec{T},$$

where V_N does not depend on the parameterization of the curve (V_N is thus called an "intrinsic" or "geometric" quantity), can be converted into the solution of

$$C_t = V_N \vec{N},$$

by a change of parameterization.

Proof. Given $C(p,t) : S^1 \times [0,T] \to \mathbb{R}^2$ as our family of curves, let $p = p(\omega, \tau)$ and $t = \tau$ with $\partial p / \partial \omega > 0$ be a reparameterization. By the chain rule

$$
\begin{aligned}
\frac{\partial}{\partial \tau} C(p(\omega, \tau), t(\omega, \tau)) &= C_p p_\tau + \partial_t C t_\tau \\
&= C_p p_\tau + \partial_t C.
\end{aligned}
$$

For the arclength parameterization s we have

$$
\begin{aligned}
C_p &= C_s s_p \\
&= \vec{T} s_p.
\end{aligned}
$$

Using these two expressions we calculate

$$
\begin{aligned}
\frac{\partial}{\partial \tau} C &= C_p p_\tau + \partial_t C \\
&= \vec{T} s_p p_\tau + V_T \vec{T} + V_N \vec{N} \\
&= (V_T + s_p p_\tau) \vec{T} + V_N \vec{N},
\end{aligned}
$$

For each fixed ω, we solve the O.D.E.

$$V_T + |C_p| p_\tau = 0,$$

and recalling the selection $t = \tau$ we arrive at

$$C_t = V_N \vec{N}. \qquad \blacksquare$$

Therefore, since our interest is the shape of the curve we can consider the "Lagrangian" form of the curve evolution:

$$C_t = \langle \vec{\mathcal{V}}, \vec{N} \rangle \vec{N}, \qquad C(s,0) = C_0(s),$$

and for $V_N = \langle \vec{\mathcal{V}}, \vec{N} \rangle$,

$$C_t = V_N \vec{N}, \qquad C(s,0) = C_0(s). \qquad (3.1)$$

3.1.1   Invariant Curve Evolution

As we have just seen, the tangential component affects only the internal parameterization and does not influence the shape of the propagating curve [67]. Thus we can consider only the normal component in the corresponding evolution equation. For example, the equi-affine invariant heat equation [174] is given by

$$C_t = \langle C_{vv}, \vec{N} \rangle \vec{N} = \kappa^{1/3} \vec{N}. \tag{3.2}$$

In [30, 156] the authors argue that the second-order derivative of a curve according to its invariant arclength leads toward the *geometric heat evolution equation* of any given group of transformation and is given by

$$C_t = C_{rr},$$

where r is an arclength defined for the specific transformation group. Actually, this statement is valid only for all subgroups of the projective group, which is the most general line-preserving group of transformations. This way it is possible [174] to reformulate the affine heat equation $C_t = C_{vv}$ with its geometric equivalent $C_t = \kappa^{1/3} \vec{N}$. In [156] it is shown that for the similarity group, given by the scaling transformation $\{\tilde{x}, \tilde{y}\}^T = \begin{pmatrix} a & 0 \\ 0 & a \end{pmatrix} \{x, y\}^T + \{b_1, b_2\}^T$, the *inverse* geometric heat equation, given by

$$C_t = -\frac{1}{\kappa} \vec{N},$$

is invariant. Yet it can be applied only to convex shapes.

Let us introduce an invariant evolution of the affine group with one given point (e.g., given origin, or any other point). Using the same argument as for the equi-affine arclength, that is areas are invariant under the affine transformation, let the simplest linear affine arclength be given by [27, 152]

$$w = \int |(C, C_p)| dp.$$

See Figure 3.4. This is an intrinsic measure that does not depend on the parameter,

$$w = \int |(C, C_p)| dp = \int \left| \left(C, \frac{C_p}{|C_p|} \right) \right| |C_p| dp = \int |(C, \vec{T})| ds,$$

where s is the Euclidean arclength.

The area of a closed shape [74] is defined by

$$\begin{aligned} \mathcal{A} &= \frac{1}{2} \oint (C, \vec{T}) ds \\ &= \frac{1}{2} \oint \langle C, \vec{N} \rangle ds. \end{aligned}$$

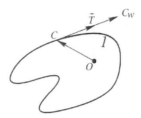

Figure 3.4: For $w =$ linear affine arclength, the area $|(C, C_w)| \equiv 1$.

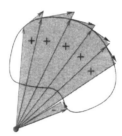

Figure 3.5: The area bounded by a closed curve is given by $\frac{1}{2} \oint (C, C_p) dp = \frac{1}{2} \oint (C, \vec{T}) ds$.

See Figure 3.5.

Therefore, the arclength (as for the equi-affine one) corresponds to an area and is thus invariant. An arclength element is given by

$$
\begin{aligned}
dw &= |(C, \vec{T})| ds \\
&= |\langle C, \vec{N} \rangle| ds.
\end{aligned}
$$

Thus,

$$
C_w = C_s \frac{ds}{dw} = C_s \frac{1}{|\langle C, \vec{N} \rangle|} = \frac{1}{|\langle C, \vec{N} \rangle|} \vec{T},
$$

and

$$
\begin{aligned}
C_{ww} &= C_{ss} \left(\frac{ds}{dw} \right)^2 + C_s \frac{d^2 s}{dw^2} \\
&= \kappa \vec{N} \left(\frac{1}{|\langle C, \vec{N} \rangle|} \right)^2 + \frac{d^2 s}{dw^2} \vec{T} \\
&= \frac{1}{\langle C, \vec{N} \rangle^2} \kappa \vec{N} + \text{ tangential component.}
\end{aligned}
$$

The corresponding linear affine heat equation given by its geometric Euclidean version is

$$C_t = \frac{1}{\langle C, \vec{N} \rangle^2} \kappa \vec{N}. \tag{3.3}$$

Observe that when we consider closed curves, the more complicated cases from the point of view of shape recognition are those in which the anchor point is located inside the curve and no tangent points are formed. In cases where tangents appear, it is possible to locate the smallest triangle that is defined by two tangent points (it is easy to prove that there exist at least two such points) and the anchor (origin) point. This triangle can be mapped into a reference triangle for every given curve; the same map brings that curve into a canonical form (see Figure 3.6). So, considering only the "interesting" cases, the geometric heat equation (3.3) is not influenced by singular values since tangent points are excluded.

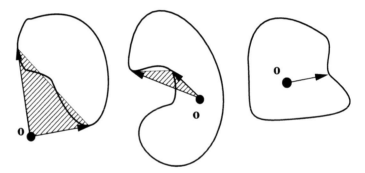

Figure 3.6: There are three possible locations for the given point: outside the shape, in which $C(p)$ is tangent to the boundary of the shape at least at two points; inside a shape, which consists of a concave part that leads again to at least two tangent points; and inside a star shape with no tangents at all, which is the interesting case from our point of view.

3.2   Properties of Curve Evolution

Before we extend and apply curve evolution-based techniques to shapes and images, we need to further explore some of their properties. The most important are the preserving of the embedding along the evolution, and convergence into a known shape.

We already have that $\vec{T}_s = \kappa \vec{N}$; another useful connection is

$$\vec{N}_s = \frac{d}{ds}\{-y_s, x_s\} = \{-y_{ss}, x_{ss}\} = -\kappa \vec{T},$$

which are the Frénet equations for planar curves. Taking a second derivative w.r.t. the arclength, we have that

$$\vec{N}_{ss} = \frac{d}{ds}(-\kappa\vec{T}) = -\kappa_s\vec{T} - \kappa^2\vec{N}.$$

Let us consider the time derivative for the general geometric curve evolution

$$\frac{\partial C}{\partial t} = V\vec{N}.$$

The length of a curve is given by integration along its arclength $\mathcal{L}(t) = \int_0^1 |C_p| dp$, so that the change of the total length in time for an evolving curve is given by

$$
\begin{aligned}
\frac{d}{dt}\mathcal{L}(t) &= \frac{\partial}{\partial t}\int_0^1 \langle C_p, C_p \rangle^{1/2} dp \\
&= \int_0^1 \frac{\langle C_{tp}, C_p \rangle}{\langle C_p, C_p \rangle^{1/2}} dp \\
&= \int_0^1 \left\langle \frac{\partial}{\partial p}(V\vec{N}), \vec{T} \right\rangle dp \\
&= \int_0^1 \left\langle |C_p|\frac{\partial}{\partial s}(V\vec{N}), \vec{T} \right\rangle dp \\
&= \int_0^1 \left\langle \frac{\partial}{\partial s}(V\vec{N}), \vec{T} \right\rangle |C_p| dp \\
&= \int_0^1 \left(\langle (V_s\vec{N}), \vec{T} \rangle - V\langle \kappa\vec{T}, \vec{T} \rangle \right) |C_p| dp \\
&= -\int_0^1 V\kappa |C_p| dp \\
&= -\int_0^L V\kappa ds,
\end{aligned}
$$

where we use integration by parts and the fact that the curve is closed. L denotes the total length of the curve.

In a similar way, we have seen that the area of a curve is given by integration of the outer product of its coordinate and its tangent along its parameter $\mathcal{A} = \frac{1}{2}\int_0^1 (C, C_p) dp$, so that the change of the total area of an evolving closed curve in time is given by

$$
\begin{aligned}
\frac{d}{dt}\mathcal{A}(t) &= \frac{d}{dt}\frac{1}{2}\int_0^1 (C, C_p) dp \\
&= \frac{1}{2}\int_0^1 \left((C_t, C_p) + (C, C_{tp}) \right) dp \\
&= \frac{1}{2}\left(\int_0^1 (V\vec{N}, C_p) dp + (C, C_t)|_0^1 - \int_0^1 (C_p, C_t) dp \right)
\end{aligned}
$$

$$
= \frac{1}{2} \left(\int_0^1 V(\vec{N}, \vec{T}) |C_p| dp - \int_0^1 (\vec{T}, V\vec{N}) |C_p| dp \right)
$$

$$
= \int_0^L V(\vec{N}, \vec{T}) ds = \int_0^L (C_t, \vec{T}) ds = \int_0^L V ds,
$$

where again we use integration by parts, change of the integration parameterization to arclength, and the fact that the curve is closed.

We can apply this kind of analysis to other geometric quantities, like the curvature. Let the "metric" along the curve be

$$
g = \langle C_p, C_p \rangle^{1/2}.
$$

Recall that $ds = gdp$ such that

$$
\frac{\partial}{\partial p} = g \frac{\partial}{\partial s}.
$$

Then, the change of the metric in time is given by

$$
\begin{aligned}
g_t &\equiv \frac{\partial}{\partial t} \langle C_p, C_p \rangle^{1/2} \\
&= \frac{\langle C_{tp}, C_p \rangle}{\langle C_p, C_p \rangle^{1/2}} \\
&= \langle \frac{\partial}{\partial p}(V\vec{N}), \vec{T} \rangle \\
&= \langle V_p \vec{N} + V \vec{N}_p, \vec{T} \rangle \\
&= V \langle -\kappa g \vec{T}, \vec{T} \rangle \\
&= -V \kappa g.
\end{aligned}
$$

The change of the curvature in time can be computed as follows:

$$
\begin{aligned}
\kappa_t &\equiv \frac{\partial}{\partial t} \left(\frac{(C_p, C_{pp})}{g^3} \right) \\
&= \frac{-3g^2 g_t (C_p, C_{pp}) + g^3 (C_{tp}, C_{pp}) + g^3 (C_p, C_{tpp})}{g^6} \\
&= \frac{3g^3 \kappa V (C_p, C_{pp}) + g^3 (\frac{\partial}{\partial p}(V\vec{N}), C_{pp}) + g^3 (C_p, \frac{\partial^2}{\partial p^2}(V\vec{N}))}{g^6} \\
&= 3\kappa^2 V + g^{-2}(\frac{\partial}{\partial s}(V\vec{N}), C_{pp}) + g^{-1}(\vec{T}, \frac{\partial^2}{\partial s \partial p}(V\vec{N})) \\
&= 3\kappa^2 V + g^{-1}(V_s \vec{N} - V\kappa \vec{T}, \frac{\partial}{\partial s} C_p) + g^{-1}(\vec{T}, \frac{\partial}{\partial s} g \frac{\partial}{\partial s}(V\vec{N})) \\
&= 3\kappa^2 V + g^{-1}(V_s \vec{N} - V\kappa \vec{T}, \frac{\partial}{\partial s}(g\vec{T})) + g^{-1}(\vec{T}, \frac{\partial}{\partial s} g(V_s \vec{N} - V\kappa \vec{T})) \\
&= 3\kappa^2 V + g^{-1}(V_s \vec{N} - V\kappa \vec{T}, g_s \vec{T} + g\kappa \vec{N}) + g^{-1}(\vec{T}, g_s V_s \vec{N} + g V_{ss} \vec{N} \\
&\quad - g V \kappa^2 \vec{N}) \\
&= 3\kappa^2 V - g^{-1} V_s g_s - V\kappa^2 + g^{-1} g_s V_s + V_{ss} - V\kappa^2 \\
&= V_{ss} + \kappa^2 V.
\end{aligned}
$$

3.2.1 The Constant Flow (Offsets)

For $V = 1$, the constant evolution equation

$$C_t = \vec{N},$$

the curvature change in time, is given by $\kappa_t = \kappa^2$ for which the solution can be found explicitly by a Riccati equation to be

$$\kappa(p,t) = \frac{\kappa(p,0)}{1 - t\kappa(p,0)}.$$

Note that singularities form at $t = 1/\kappa(p,0)$ as the denominator approaches zero; see Figure 3.7.

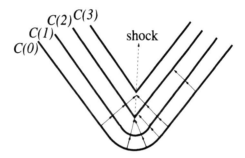

Figure 3.7: Singularities in the curvature are formed at $t = 1/\kappa(p,0)$. As the curve continues its constant flow, the normals will continue to collide at "shock points" and form a trajectory known as the medial axis, symmetry set, or skeleton of the initial shape of the closed curve.

The change in length for constant evolution is given by $\mathcal{L}_t = -\int_0^L V ds = -\int_0^{2\pi} d\theta = -2\pi$, which means that the length is reduced in a constant rate while shocks (singularities in the curvature) do not occur in the propagating curve. This means that a circle with a given length $\mathcal{L}(0) = 2\pi r$ will vanish at $t = \mathcal{L}(0)/(2\pi) = r$. The change of area for this evolution is given by $\mathcal{A}_t = \int_0^L (\vec{N}, \vec{T}) ds = -\int_0^L ds = -L$, which means that the rate of change in area depends on the current length of the curve.

3.2.2 The Curvature Flow

For the geometric heat equation $C_t = C_{ss} = \kappa\vec{N}$, we have that the change in curvature is given by $\kappa_t = \kappa_{ss} + \kappa^3$, which is a reaction diffusion geometric question. The change in length for this evolution is given by $\mathcal{L}_t = -\int_0^L \kappa^2 ds$. The change of area for the curvature flow is given by $\mathcal{A}_t = \int_0^L (\kappa\vec{N}, \vec{T}) ds = -\int_0^L \kappa ds = -2\pi$, which means that the rate of change in area is a constant. We readily have that $\mathcal{A}(t) = \mathcal{A}(0) - 2\pi t$, which means that the curve will reduce its area in linear time and if it exists along the evolution it will vanish at $t = \mathcal{A}(0)/(2\pi)$.

A powerful property for the curvature flow comes from the following two theorems:

Theorem 1 [Gage–Hamilton] [74]: *If C is a convex curve embedded in the plane $I\!R^2$, the curvature flow shrinks C to a point. The curve remains convex and becomes circular as it shrinks in the sense that*
a. *the ratio of the inscribed radius to the circumscribed radius approaches 1, and*
b. *the ratio of the maximum curvature to the minimum curvature approaches 1, and*
c. *the higher-order derivatives of the curvature converge to 0 uniformly.*

This result is followed by an even more powerful theorem:

Theorem 2 [Grayson] [82]: *Let $C(\cdot, 0) : S^1 \to I\!R^2$ be a smooth embedded curve in the plane. Then $C : S^1 \times [0, T) \to I\!R^2$ exists satisfying $C_t = \kappa \vec{N}$, the curvature flow. $C(\cdot, t)$ is smooth for all t, it converges to a point as $t \to T$, and its limiting shape as $t \to T$ is a round circle, with convergence in the C^∞-norm.*

The Grayson theorem proves that the curvature flow takes any simple non-intersecting curve into a convex one, and then by the Gage and Hamilton theorem it converges to a circular point. See Figure 3.8. In a similar way, Sapiro and Tannenbaum [174, 181] show an analog behavior for the equi-affine invariant flow, where for the equi-affine case, the convergence is to an ellipse as expected.

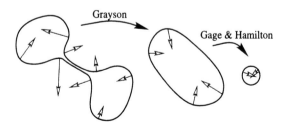

Figure 3.8: The Grayson theorem states that the curvature flow evolves any planar curve into a circular point in finite time.

3.3   Surface Evolution

One of the most important results in differential geometry is Gauss' "Theorema Egregium," which states that *The Gaussian curvature K of a surface is invariant under local isometries.* It means that the Gaussian curvature does not depend on the space that contains the surface, the embedding space. Apparently, some other geometric measures share the same intrinsic property. One example is the geodesic curvature, κ_g, of a curve embedded in a surface. If we paint a curve on a surface, and band the surface without stretching or tearing it apart, then the geodesic curvature of the curve on the bended surface is the same as it was before the surface was bended. There are promising indications of nice behavior of curves on surfaces that flow according to their geodesic curvature. Obviously, closed curves painted on bended planes will vanish at a point as they flow by their geodesic curvature, since this flow is identical to the curvature flow of the curve on a plane.

Recall (see Chapter 1, Ex. 6) that the first variation of the area of a surface yields the Euler–Lagrange equations $H\vec{N} = 0$. The mean curvature surface flow given by

$$\mathcal{S}_t = H\vec{N}$$

is the steepest descent flow toward a minimal surface and therefore an efficient way to reduce the area of a given surface. It is a natural extension of the curvature flow for planar curves. However, the most powerful geometrical property we had for curves, the preservation of the topology along the evolution, is lost for surfaces. An interesting open question is if there exists a local geometric evolution rule that preserves the topology of surfaces in 3D.

In a similar way to the affine invariant flow for curves, the equi-affine invariant flow for surfaces is given by

$$\mathcal{S}_t = K^{1/4}\vec{N}.$$

Yet again, unfortunately the surface topology is not preserved by this flow.

3.3.1   Images That Flow as Surfaces

A minimal surface is the surface with the least area that satisfies given boundary conditions. It has nice geometrical properties and is often used as a natural model of various physical phenomena, for example, soap bubbles, "Plateau's problem," in computer aided design, in architecture (structural design), and even for medical imaging [37, 36]. J.L. Lagrange realized in 1762 [126] that the mean curvature equal to zero is the requirement for area minimization. Hence, the mean curvature flow is the most efficient flow toward a minimal surface. Numerical schemes for the mean curvature flow, and the construction of minimal surfaces under constraints, were introduced at the beginning of the modern age of numerical analysis [49] and are still the subject of ongoing numerical research [43, 44, 40].

For constructing the mean curvature flow of the image as a surface, we follow three steps:

1. Given the surface S that evolves according to the geometric flow

$$S_t = \vec{F},$$

where $\vec{F}$ is an arbitrary smooth flow field. The geometric deformation of S may be equivalently written as

$$S_t = \langle \vec{F}, \vec{N} \rangle \vec{N}, \tag{3.4}$$

where $\vec{N}$ is the unit normal of the surface at each point, and $\langle \vec{F}, \vec{N} \rangle$ is the inner product (the projection of $\vec{F}$ on $\vec{N}$). The tangential component affects only the internal parameterization of the evolving surface and does not influence its geometric shape.

2. The mean curvature flow is given by

$$S_t = H\vec{N},$$

where H is the mean curvature of S at every point. Let us use the relation given in Step 1:

3. Consider the image function $I(x, y)$ as a parameterized surface $S = (x, y, I(x, y))$. We may write the mean curvature flow as

$$S_t = \frac{H}{\langle \vec{N}, \vec{Z} \rangle} \vec{Z},$$

for any smooth vector field $\vec{Z}$ defined on the surface. Especially, we may choose $\vec{Z}$ as the $\hat{I}$ direction, that is, $\vec{Z} = (0, 0, 1)$. In this case

$$\frac{1}{\langle \vec{N}, \vec{Z} \rangle} \cdot \vec{Z} = \sqrt{1 + I_x^2 + I_y^2} \cdot (0, 0, 1).$$

Fixing the (x, y) parameterization along the flow (i.e., using the fixed x, y-plane as the natural parameterization), we have $S_t = \frac{\partial}{\partial t}(x, y, I(x, y)) = (0, 0, I_t(x, y))$. Thus, for tracking the evolving surface, it is enough to evolve I via

$$I_t = H\sqrt{1 + I_x^2 + I_y^2},$$

where the mean curvature H is given as a function of the image I (see Figure 3.9), and

$$H = \text{div} \left(\frac{\nabla I}{\sqrt{1 + |\nabla I|^2}} \right) = \frac{(1 + I_y^2)I_{xx} - 2I_xI_yI_{xy} + (1 + I_x^2)I_{yy}}{(1 + I_x^2 + I_y^2)^{3/2}}.$$

See [42, 43] for the derivation of H (as D.L. Chopp summarizes the original derivation by J.L. Lagrange from 1762).

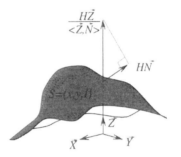

Figure 3.9: Consider the surface mean curvature flow $\mathcal{S}_t = H\vec{N}$, mean curvature H in the surface normal direction $\vec{N}$. A geometrically equivalent flow is the flow $\partial(x, y, I)/\partial t = H(1 + |\nabla I|^2)^{1/2} \cdot (0, 0, 1)$, which yields the mean curvature flow when projected onto the normal.

We end up with the following evolution equation:

$$I_t = \frac{(1 + I_y^2)I_{xx} - 2I_xI_yI_{xy} + (1 + I_x^2)I_{yy}}{1 + I_x^2 + I_y^2}, \tag{3.5}$$

with the image itself as initial condition $I(x, y, 0) = I(x, y)$.

In Chapter 10 we introduce a compact notation for Laplacian on curved surfaces, known as the *Laplace–Beltrami operator*. Using this second-order operator, Δ_g, and the metric g, Eq. (3.5) may be read as

$$I_t = g\Delta_g I,$$

while the Beltrami flow (selective mean curvature flow) $I_t = \Delta_g I$ is given explicitly for the simple 2D case as

$$I_t = \frac{(1 + I_y^2)I_{xx} - 2I_xI_yI_{xy} + (1 + I_x^2)I_{yy}}{(1 + I_x^2 + I_y^2)^2}; \tag{3.6}$$

see Figure 3.10.

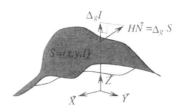

Figure 3.10: Consider the mean curvature H in the surface normal direction $\vec{N}$. It can also be expressed as $H\vec{N} = \Delta_g S$. The Beltrami operator that operates on I: $\Delta_g I$ is the third component of this vector: projection onto the I $(\vec{Z})$ direction.

3.4 Exercises

1. Implement curve evolution via constant velocity $C_t = \vec{N}$ and the curvature flow $C_t = \kappa \vec{N}$. Approximate the initial curve by a set of points (Heron's formula from Exercise 7 in Chapter 1 can be used to approximate the curvature). Apply your procedure to convex and nonconvex curves and report your results and the numerical difficulties.

2. Compute the geometric heat flow for the shear transformation, based on your solution to Exercise 2.3. It is enough to use only the projection of the invariant curvature vector onto the Euclidean normal.

3. Compute $\vec{T}_t, \vec{N}_t, \theta_t, \kappa_t$ for the curvature flow, the affine flow, and the constant flow [74, 187, 99, 174].

4. Consider the time derivative for the general geometric curve evolution

$$\frac{\partial C}{\partial t} = V \vec{N}.$$

Prove the following lemma that can help us compute the changes of the geometric quantities of an evolving curve.

Lemma 3 *For a curve evolving by $C_t = V \vec{N}$, we have*

$$\frac{\partial}{\partial t}\frac{\partial}{\partial s} = \frac{\partial}{\partial s}\frac{\partial}{\partial t} + V\kappa\frac{\partial}{\partial s}$$

and specifically [74] for $V = \kappa$

$$\frac{\partial}{\partial t}\frac{\partial}{\partial s} = \frac{\partial}{\partial s}\frac{\partial}{\partial t} + \kappa^2\frac{\partial}{\partial s}.$$

5. A small research project: Find a bound on the location of the vanishing point for a curve evolving via the geometric heat equation $C_t = C_{ss}$.

6. Compute the surface normal for the graph parameterized surface $\{x, y, z(x, y)\}$.

7. **a.** Consider the flow

$$z_t = \frac{z_{xx}(1 + z_y^2) - 2z_x z_y z_{xy} + z_{yy}(1 + z_x^2)}{(1 + z_x^2 + z_y^2)^{k/2}},$$

for $k \in \{2, 3, 4\}$. For which k is the flow geometrically equivalent, in the sense of Lemma 2, to the mean curvature flow of the surface $\{x, y, z(x, y)\}$? Explain the relation to the result of Exercise 6 in Chapter 1.
b. Apply the above flow to a gray-level image $z(x, y; 0) = I(x, y)$. Use periodic boundary conditions, central derivatives in space,

$$z_x \approx \frac{z_{i+1,j} - z_{i-1,j}}{2\Delta x}$$

$$z_{xx} \approx \frac{z_{i+1,j} - 2z_{i,j} + z_{i-1,j}}{\Delta x^2}$$

$$z_{xy} \approx \frac{z_{i+1,j+1} - z_{i-1,j+1} - z_{i+1,j-1} + z_{i-1,j-1}}{4\Delta x^2},$$

and forward in time. Compare the results of your flows for $k = 2, 3, 4$.

8. Research project:
   a. Prove the median scheme for curvature computation [85].
   b. Prove the Catté–Dibos $(IS + SI)_{r\theta}$ scheme; see [85].
   c. Extend these morphological methods to find the affine curvature μ, or C_{vv} projection onto the normal: $\kappa^{1/3}$.

4

The Osher–Sethian Level Set Method

In this chapter we introduce the level set approach for curve and surface evolution. One way to represent a curve is as a level set or an equal-height contour of a given function. The intersection between this function and a plane parallel to the coordinate plane yields the curve. This function is an *implicit representation* of its level set, and actually of all its level set curves. This obvious relation between explicit curves and their implicit representation was used by Osher and Sethian [161] to introduce a powerful way for numerically tracking evolving interfaces; see also [155].

Consider the planar curve evolution $C_t = V_N \vec{N}$. Let $\phi(x, y; t)$ be an implicit representation of the curve so that $C(s, t) = \{(x, y) | \phi(x, y; t) = 0\}$, that is, the zero level set of a time-varying surface function $\phi(x, y; t)$. Then, the propagation rule for ϕ that yields the correct curve propagation equation is given by [161]

$$\frac{\partial \phi}{\partial t} = V_N |\nabla \phi| \quad \text{given} \quad \phi^{-1}(0) = C(0);$$

we will prove this relation in the next section. In some problems it is natural to use a given image I as initialization for the implicit function $\phi(x, y; 0) = I$. The level set formulation for curve evolution thus provides a beautiful link between dynamic curves and scalar (gray-level) evolving images.

Tracking the *zero level set* of the bivariate function $\phi(x, y)$ propagating in time overcomes numerical and topological problems of the propagating level set curve in an elegant way. The implicit formulation for the implementation of propagating curves was explored by Osher and Sethian [161]. Sethian called it the *Eulerian formulation* for curve evolution [189].

4.1 The Eulerian Formulation

The Osher–Sethian level set formulation allows the developments of efficient and stable numerical schemes in which topological changes of the propagating curve are automatically handled.

While implementing the evolution given by the explicit Lagrangian formulation, $C_t = V_N \vec{N}$, one should handle topological changes in the evolving curve by external procedures. Such a procedure should monitor the process and detect possible mergings and splittings of the curve. It was also shown [187, 161, 189] that such implementations are very sensitive to the formation of high curvature and sharp corners. Even an initial smooth curve can develop curvature singularities. The question is how to continue the evolution after singularities appear. The natural way is to choose the solution that agrees with the *Huygens principle*. Viewing the curve as the front of a burning flame, this solution states that *once a particle is burned, it cannot be re-ignited*. It can also be proved that from all the *weak* solutions of the explicit Lagrangian formulation $C_t = V_N \vec{N}$, the one derived from the Huygens principle is unique and can be obtained by a constraint denoted as the "entropy condition for curve evolution [187]." We will elaborate on this topic through examples.

Let $\phi(x, y; t) : \mathbb{R}^2 \times [0, T) \rightarrow \mathbb{R}$ be an implicit representation of the curve $C(s, t)$, so that the zero level set $\phi(x, y; t) = 0$ is the set of points constructing the curve $C(s, t)$. In other words, the trace of the curve C at time t is given by the zero level set of the function ϕ at time t:

$$C(t) = \phi(\cdot, \cdot; t)^{-1}(0).$$

The demand of C being the zero level set is arbitrary, and actually any other level set may serve the same purpose. See Figure 4.1. The problem is how to evolve the function ϕ in time so that its zero level set tracks the time-varying curve $C(t)$.

Denote by $\nabla \equiv (\partial/\partial x, \partial/\partial y)$ the gradient operator. Then, from basic calculus, we have

Lemma 4 *The planar unit normal of the curve $C = \phi^{-1}(c)$, where c is an arbitrary constant selecting the level set, is given by $\vec{N} = \pm \nabla \phi / |\nabla \phi|$.*

Proof. Let s be the arclength parameter of C. Then, along the equal-height contour C the change of ϕ is zero:

$$\phi_s = 0 = \phi_x x_s + \phi_y y_s.$$

This expression $\langle \nabla \phi, C_s \rangle = 0$, determines that $\nabla \phi$ is orthogonal to $C_s = \vec{T}$.
∎

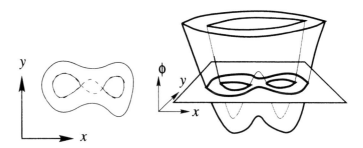

Figure 4.1: Left: A simple curve evolves inwards and splits into two simple curves. Right: The curves embedded as level sets of an evolving higher-dimensional function that does not change its topology.

By the level set motion we have

$$\phi(x + x_t \Delta t, y + y_t \Delta t, t + \Delta t) = \phi(x, y, t).$$

Taylor expansion about (x, y, t) yields

$$\phi(x + \Delta t x_t, y + \Delta t y_t, t + \Delta t) = \phi(x, y, t) + \Delta t x_t \phi_x(x, y, t)$$
$$+ \Delta t y_t \phi_y(x, y, t) + \Delta t \phi_t(x, y, t) + O(\Delta t^2).$$

Plugging the level set motion identity into the above expression and dividing by Δt we have

$$0 = x_t \phi_x(x, y, t) + y_t \phi_y(x, y, t) + \phi_t(x, y, t) + O(\Delta t).$$

We take $\Delta t \to 0$ and get

$$\phi_t = -(\phi_x x_t + \phi_y y_t).$$

Note that all we did is to apply the chain rule along the evolving curve $\phi(x(t), y(t), t) =$ constant. That is, $(d/dt)\phi(x(t), y(t); t) = 0$, which yields $\phi_t + \phi_x x_t + \phi_y y_t = 0$. This equation may be written as

$$\begin{aligned}
\phi_t &= -\langle \nabla \phi, C_t \rangle \\
&= -\langle \nabla \phi, V_N \vec{N} \rangle \\
&= -\left\langle \nabla \phi, V_N \frac{\nabla \phi}{|\nabla \phi|} \right\rangle \\
&= -V_N \left\langle \nabla \phi, \frac{\nabla \phi}{|\nabla \phi|} \right\rangle \\
&= -V_N |\nabla \phi|,
\end{aligned}$$

which is the Eulerian formulation for curve evolution. Given **any** smooth function $\phi_0(x, y)$ such that $\phi_0^{-1}(0) = C_0$, we can rewrite the last result as

$$\phi_t = -V_N |\nabla \phi|, \qquad \phi(x, y, 0) = \phi_0(x, y), \qquad (4.1)$$

which is a Hamilton–Jacobi type of equation. This formulation of planar curve evolution processes frees us from the need to take care of the possible topological changes in the propagating curve. Sethian [189] named the above *Eulerian formulation* for front propagation, because it is written in terms of a fixed coordinate system. Note that the sign for the Eulerian formulation depends on the selection of the normal direction. For instance, for $\vec{N} = -\nabla\phi/|\nabla\phi|$ we have $\phi_t = V_n|\nabla\phi|$.

Let us give a geometric interpretation for the Osher–Sethian level set formulation. We compute the differential change in $\phi(x, y; t)$ that corresponds to the normal velocity of its zero level set at a specific point $(x_0, y_0, 0)$; see Figure 4.2. For that we use a first-order approximation of $\phi(x, y; t)$ at the point $(x, y, t) = (x_0, y_0, 0)$, and assume w.l.o.g. that $\vec{N}$ aligns with the x-axis. On the left, the level sets of the function $\phi(x, y; 0)$ are shown as thin curves, while the zero level set is painted as a thick curve. The right frame presents a cross section of ϕ at $t = 0$ and $t = \Delta t$. We readily have by similarity of triangles that

$$\frac{\Delta\phi}{\Delta x} = \frac{\phi(x_0, \Delta t) - \phi(x_0, 0)}{V_N \Delta t}.$$

As we take the limits $\Delta t, \Delta x \to 0$ we obtain again the differential equation $\phi_t = V_N|\nabla\phi|$. We see that in order for the zero set to move by $V_N dt$ we need to locally raise ϕ by $|\frac{d\phi}{dx}|V_N dt$.

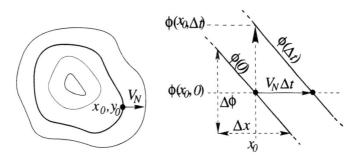

Figure 4.2: A geometric interpretation for the change in ϕ at (x_0, y_0).

The normal component V_N can be any smooth scalar function. An important observation is that any geometric property of the curve C can be computed from its implicit representation ϕ. The curvature, for example, plays an important role in many applications.

Lemma 5 *The curvature κ of the planar curve $C = \phi^{-1}(c)$ is given by*

$$\kappa = -\frac{\phi_{xx}\phi_y^2 - 2\phi_x\phi_y\phi_{xy} + \phi_{yy}\phi_x^2}{(\phi_x^2 + \phi_y^2)^{3/2}}. \tag{4.2}$$

Proof. Along C, the function ϕ does not change its values. Therefore, $\partial^n \phi / \partial s^n = 0$, for any n. Particularly, for $n = 2$,

$$
\begin{aligned}
0 &= \frac{\partial^2 \phi}{\partial s^2} \\
&= \frac{\partial}{\partial s}(\phi_x x_s + \phi_y y_s) \\
&= \phi_{xx} x_s^2 + 2\phi_{xy} x_s y_s + \phi_{yy} y_s^2 + \phi_x x_{ss} + \phi_y y_{ss} \\
&= \phi_{xx} x_s^2 + 2\phi_{xy} x_s y_s + \phi_{yy} y_s^2 + \langle \nabla\phi, C_{ss} \rangle.
\end{aligned}
\tag{4.3}
$$

Recall that $\vec{N} = \{-y_s, x_s\} = \nabla\phi / |\nabla\phi|$, and that by definition $C_{ss} = \{x_{ss}, y_{ss}\} = \kappa \vec{N}$. Or explicitly

$$
\begin{cases}
y_s &= -\dfrac{\phi_x}{\sqrt{\phi_x^2 + \phi_y^2}}, \\[4ex]
x_s &= \dfrac{\phi_y}{\sqrt{\phi_x^2 + \phi_y^2}},
\end{cases}
$$

and

$$
\begin{cases}
x_{ss} &= \kappa \dfrac{\phi_x}{\sqrt{\phi_x^2 + \phi_y^2}}, \\[4ex]
y_{ss} &= \kappa \dfrac{\phi_y}{\sqrt{\phi_x^2 + \phi_y^2}}.
\end{cases}
$$

Introducing these two expressions into Eq. (4.3), we conclude that

$$
\begin{aligned}
0 &= \frac{\phi_{xx}\phi_y^2 - 2\phi_x \phi_y \phi_{xy} + \phi_{yy}\phi_x^2}{|\nabla\phi|^2} + \langle \nabla\phi, C_{ss} \rangle \\
&= \frac{\phi_{xx}\phi_y^2 - 2\phi_x \phi_y \phi_{xy} + \phi_{yy}\phi_x^2}{|\nabla\phi|^2} + |\nabla\phi|\kappa. \qquad \blacksquare
\end{aligned}
$$

The same equation for the curvature of the level set of ϕ may be equivalently written as

$$
\kappa = -\mathrm{div}\left(\frac{\nabla\phi}{|\nabla\phi|} \right).
$$

This equation applies to the mean curvature computation of the level set in any dimension.

4.2   From Curve to Image Evolution

Scale space in image analysis is the idea of "flowing" an intensity image so as to simplify its structure and make the segmentation/separation of objects and the recognition tasks easier. One could consider it as a way of getting rid of noise without destroying the valuable information as a first step, and then as a smoothing technique that keeps the important features "alive" while smoothing away the noise. The simplest example is the heat operator operating on a gray-level image.

In the following chapters we will learn about the geometric framework for image processing in which images are considered as surfaces [198, 118, 117, 107]. We will justify the idea of images as surfaces rather than gray-level sets, especially when such are not natural like color images. Let us start with the philosophy of images as a collection of gray-level sets. When we deal with invariance, we need some simplifications that would make our mathematical goal plausible. We thus consider the world to be flat, like observing pictures in a museum, reduce the effects of perspective when possible, etc. See Figure 4.3. Then, each gray-level set "lives" on a plane. It is obviously an inaccurate and oversimplified assumption of the world and our perception. However, in many cases this is exactly what we want. One example is postprocessing after texture mapping in computer graphics.

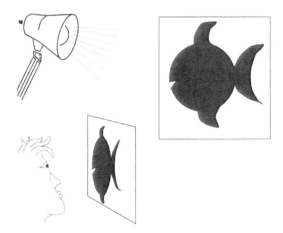

Figure 4.3: A simplified perception model of a flat shape is the affine transformation of the shape.

In computer graphics, after a texture is mapped onto a surface in 3D, this surface may be bent, then we render the surface and present it on a flat screen. The question is how can one process the image on the surface such that the processing is invariant to the deformation and bending of the surface? We want our flow to be "bending invariant" or, in other words, invariant to isometric mapping of the surface.

We start with a brief introduction to the "evolution" of scale space and the linear scale space of images as considered by Witkin in [216]. We conclude with the geodesic scale space and put it in perspective with the rest of the geometric invariant flows.

When considering scale space of images, we traditionally refer to the linear heat equation $I_t = \Delta I$. Being linear, it has the nice property of result prediction via a simple convolution with the right kernel, as we showed in Chapter 1. For results after a long duration of the "flow" we can use the Fourier domain, with a simple multiplication and low computational complexity.

However, the "topological complexity" of the image may increase in the following sense. If we track a single equal-height contour (level set) of the gray level, we notice that a simple connected contour may split into two. See Figure 4.4. This and other reasons lead to the consideration of geometric scale spaces.

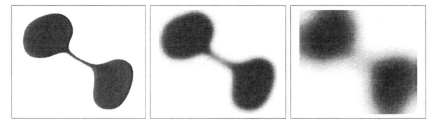

Figure 4.4: The linear scale space may change the topology of the gray-level sets.

The most natural selection is the curve-shortening flow of the level sets. With the simple and nice connection by Osher and Sethian level set formulation, it leads to image evolution, where the image I is used as the implicit function ϕ; see [68] for a rigorous support for this implicit evolution. Recall that by Grayson's theorem [82] a curve that flows according to this flow first becomes convex, and then disappears at a circular point, while embedding is preserved.

Next, the question raised by two different groups [3, 180] was how to produce an affine invariant flow for images. The simplest answer turned out to be the equi-affine invariant flow that is based on the equi-affine curve evolution. In the following chapters we will learn about a framework that captures the geometric- and the variational-based flows as a result of applying Beltrami operator w.r.t. the relevant metric on the image.

But first, let us introduce a flow for images of painted surfaces that is invariant to "bending" of the surface. For that we use the fact that geodesic curvature flow of the gray-level sets, grouped together with some nice properties introduced by Grayson in [83], lead to a flow of the whole image. Our goal now is to explore a bending invariant flow—a flow that is invariant to *isometric mapping* or *bending* of the surface. See Figure 4.5. The flow depends on the spatial derivatives of the surface and the image

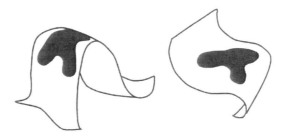

Figure 4.5: Isometric mapping or bending of a painted surface.

but does not require the mapping itself. We say that this flow is invariant to the bending process.

What are the desirable properties of such a flow? (1) It should obviously be invariant to surface bending. (2) The gray-level sets embedding should be preserved along the flow. (3) The level sets should exist for all evolution time, disappear at points or converge to geodesics. (4) The image topology should be simplified. (5) We would like the scale space to be a shortening flow of the image level sets. (6) Its numerical implementation should be simple, consistent, and stable.

In [94] it is shown that the geodesic curvature flow can serve this purpose. As we have seen, the flow

$$C_t = \kappa_g \mathcal{N}$$

is based on an intrinsic quantity, the geodesic curvature vector, and thus the flow itself is intrinsic for curves on surface. The idea to use the geodesic curvature vector as a dynamic flow field to simplify curves on the surface was studied by Grayson [83] and used in [44, 43, 109, 104] to refine curves on surfaces into geodesics.

4.2.1   Implementation

Consider the gray-level image, and pick a specific level set of this image. Project the flow of the curve onto the image coordinate plane, and use the Eulerian formulation for that curve via the level set connection. The

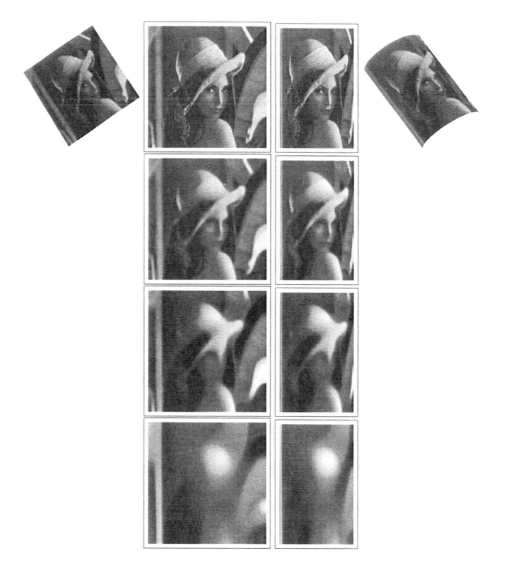

Figure 4.6: Geodesic curvature flow for an image on a flat plane, and the same image on a cylinder.

result is an invariant flow for the whole image. Technically, we follow the three following steps: First express the flow as a function of a curve and the surface derivatives in 3D. Next, project the evolution and find the corresponding flow that describes its projection on the image coordinates plane. It enables us to implement the flow on the image plane rather than in the 3D space. Last, we embed the curve as a level set in a higher-dimensional function, which is the image itself, and obtain the desired flow. Geometrically, we project the geodesic curvature vector onto the image plane, and then consider only the normal component; recall that the other component affects only the parameterization. An interesting observation is that for cylinders, we have all the nice properties that we used to have for the curvature flow of curves in the plane. See Figure 4.6.

Next, comes the interesting question: What should be done when we deal with color images? Here, the nice properties of curves and level sets are not relevant anymore. Our answer will be motivated by the observation that level sets that flow by their curvature may be viewed as the limit case of a surface with high amplitude that evolves according to its mean curvature so as to minimize its area. The question then is how to produce a geometric measure for color images that will be geometric, incorporate the nature of our color perception, and enable us to selectively smooth, enhance, and maybe even sharpen many types of images. We will address all these issues in the following chapters. See [12] for examples of image flow on implicit surfaces, and [198, 108] for image flow on an explicit representation of the surface.

4.3 Exercises

1. **a.** Find the level set expression of the Euclidean invariant geometric flow for gray-level images.
   b. Find the level set expression of the equi-affine invariant flow for gray-level images.
   c. Implement the above flows and verify numerically the invariant properties.

2. Derive the level set formulation for the general curve flow $C_t = \vec{V}$, where $\vec{V}$ is a general velocity that does not necessarily coincide with the normal direction.

3. Prove that the geodesic curvature flow for the level sets of the image $I(x, y)$ painted on the surface $\{x, y, z(x, y)\}$ is given by

$$
\begin{aligned}
I_t = \; & \big((I_x^2 I_{yy} - 2 I_x I_y I_{xy} + I_y^2 I_{xx})(1 + |\nabla z|^2) \\
& + \langle \nabla z, \nabla I \rangle (z_{xx} I_y^2 - 2 I_x I_y z_{xy} + z_{yy} I_x^2) \big) \, / \\
& \big((1 + |\nabla z|^2)(I_x^2(1 + z_y^2) + (I_y^2(1 + z_x^2) - 2 z_x z_y I_x I_y) \big).
\end{aligned}
$$

Assume that each gray-level set of I is evolving according to its geodesic curvature vector $\kappa_g \mathcal{N}$ on the surface $z(x, y)$, where x, y are the image plane coordinates.

4. A distance map $\phi(x, y)$ of a contour C assigns a positive distance from the contour to each point outside the contour, and a negative distance from the contour to each point in the interior of the contour. Thereby, a distance map of a circle is a cone in the x, y, ϕ space. The map ϕ can serve as a level set representation of C. Show that for a distance map implicit representation ϕ of a closed contour C, the level set curvature is $\kappa = \Delta\phi$.

   Construct a numerical scheme for the curvature flow of a curve C based on this observation.

5. Compute the implicit form of κ_s. That is, $\kappa_s = F(\phi_x, \phi_{xx}, ...)$, for $C = \{(x, y)|\phi(x, y) = 0\}$.

6. **a.** Given the curve evolution $C_t = \beta(\kappa)\vec{N}$, let the curve $C(p, 0)$ be the zero set of its distance map $\phi(x, y; 0)$. What is the evolution equation for ϕ such that it is kept a distance function of its zero level set along the evolution?
   b. Next, consider the more general curve evolution $C_t = \beta(x, y)\vec{N}$. What is the evolution for a distance map, ϕ, such that the distance map property from the zero set is preserved along the evolution? Hint: Design the evolution for ϕ such that each point moves with the same velocity as that of its corresponding closest zero level set point.

7. Implement a contour finder for a given function $\phi : \mathbb{R}^2 \to \mathbb{R}$ given as its samples on a discrete grid $\phi_{ij} = \phi(i\Delta x, j\Delta y)$. Write a procedure that processes only four values $\{\phi_{ij}, \phi_{i+1,j}, \phi_{i+1,j+1}, \phi_{i,j+1}\}$ at a time and returns a segment of the desired contour. What are the possible ambiguities of your procedure? Suggest a solution for these ambiguities. Optional: Extend your method to 3D, that is, design and program a "marching cube" algorithm.

5

The Level Set Method: Numerical Considerations

Numerical analysis of conservation laws plays an important role in the implementation of curve evolution equations. This chapter reviews the relevant basic concepts in numerical analysis and the relation between curve evolution, Hamilton–Jacobi partial differential equations, and differential conservation laws. This relation enabled Osher and Sethian [161] to introduce finite difference approximations, based on the theory of conservation laws, into curve evolution.

5.1 Finite Difference Approximation

Continuous-case analysis is obviously important when analyzing PDEs. However, although accurate analysis serves an important role in understanding the behavior of the equation, when implementing a numerical approximation of such an equation on a digital computer, one must address several other topics as well.

An example of a simple, yet very important, question is how to approximate $u_x(x)$, the first derivative of the function $u(x) : \mathbb{R} \to \mathbb{R}$. Assume that $u(x)$ is sampled by uniform samples of its values at equal intervals of Δx. Denote u_i to be its ith sample, that is, $u_i \equiv u(i\Delta x)$, and $D^x u_i$ as the *finite difference approximation* of the function u at the point $x = i\Delta x$. In approximating u_x one should consider computation efficiency, accuracy, and consistency with the continuous case. The first approximation is the *centered* difference finite approximation, given by

$$D^x u_i \equiv \frac{u_{i+1} - u_{i-1}}{2\Delta x}.$$

In order to measure the quality of this numerical approximation we define the *local truncation error*. It is defined by replacing the approximated solution in the finite difference method by the true solution $u(j\Delta x, n\Delta t)$. Let us replace, for example, u_j^{n+1} by the Taylor series about $u(x,t)$, namely $u + \Delta t u_t + (1/2)\Delta t^2 u_{tt} + \cdots$. We do the same for the spatial derivatives, and arrive at the error bound that is a function of Δx and Δt. A first-order accurate scheme is a differential method with local truncation error (for $\Delta t/\Delta x = $ constant) of $O(\Delta t)$ (as $\Delta t \to 0$).

Let us denote $h = \Delta x$ for simplicity of notation. Then the Taylor expansion of u_{i+1} about u_i is given by

$$u_{i+1} = u(ih + h) = u(ih) + hu'(ih) + \frac{1}{2!}h^2 u''(ih) + \cdots$$
$$+ \frac{1}{n!}h^n u^{(n)}(ih) + \cdots$$

Similarly,

$$u_{i-1} = u(ih - h) = u(ih) - hu'(ih) + \frac{1}{2!}h^2 u''(ih) + \cdots$$
$$+ \frac{1}{n!}(-h)^n u^{(n)}(ih) + \cdots$$

Such that we have

$$u_{i+1} - u_{i-1} = u(ih) + hu'(ih) + \frac{1}{2!}h^2 u''(ih) + \cdots$$
$$+ \frac{1}{n!}h^n u^{(n)}(ih) + \cdots - \left(u(ih) - hu'(ih) + \frac{1}{2!}h^2 u''(ih) \right.$$
$$\left. + \cdots + (-1)^n \frac{1}{n!}h^n u^{(n)}(ih) + \cdots \right)$$
$$= 2hu'(ih) + 2\frac{1}{3!}h^3 u'''(ih) + O(h^4).$$

Thereby, the central numerical approximation for u_x is given by

$$D^x u_i = \frac{u_{i+1} - u_{i-1}}{2h} = u'(ih) + \frac{1}{3!}h^2 u'''(ih) + O(h^3) = u'(ih) + O(h^2).$$

Thus the truncation error is of $O(\Delta x^2)$.

The *forward* finite approximation is defined as

$$D_+^x u_i \equiv \frac{u_{i+1} - u_i}{\Delta x},$$

and the *backward* approximation is

$$D_-^x u_i \equiv \frac{u_i - u_{i-1}}{\Delta x}.$$

Applying the Taylor expansion about u_i it is easy to show that in both the forward and backward approximations the truncation error is of $O(\Delta x)$.

As $\Delta x \to 0$, the derivative approximations we just introduced converges to the continuous case for smooth functions. Such convergence is referred to as *consistency* with the continuous case.

5.2   Conservation Laws and Hamilton–Jacobi Equations

The curve evolution equations are differential rules describing the change of the curve, or its evolution, in "time." As we have seen, the level set formulation puts curve evolution equations into the level set formulation that has the flavor of conservation laws. Following [128]: A conservation law asserts that the rate of change of the *total amount* of substance contained in a fixed domain G is equal to the *flux* of that substance across the boundary of G. Denoting the *density* of that substance by u, and the flux by $\vec{f}$, the conservation law is

$$\frac{d}{dt}\int_G u\,dx = -\int_{\partial G}\langle \vec{f},\vec{n}\rangle dS,$$

where $\vec{n}$ denotes the outward normal to G and dS the surface element on ∂G, which is the boundary of G, so that the integral on the right measures the outflow—hence the minus sign. See Figure 5.1. Applying the divergence theorem, taking d/dt under the integral sign, dividing by the volume of G, and shrinking G to a point where all partial derivatives of u and $\vec{f}$ are continuous, we obtain the *differential conservation law*:

$$u_t + \operatorname{div}\vec{f} = 0.$$

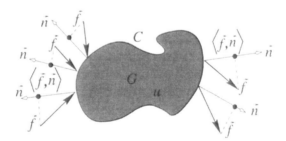

Figure 5.1: The change of u in G is an integral of the projection of the flow $\vec{f}$ along the boundary C on the boundary normal $\vec{n}$.

Consider the simple one-dimensional case in which the integral (by x and t) version of a conservation law has the explicit form of

$$\int_{x_0}^{x_1}(u(x,t_1)-u(x,t_0))dx + \int_{t_0}^{t_1}(f(x_1,t)-f(x_0,t))dt = 0.$$

A solution u is called a *generalized solution* of the conservation law if it satisfies the above integral form for every interval (x_0,x_1) and every time interval (t_0,t_1). Taking $x_1 \to x_0$, $t_1 \to t_0$, and dividing by the volume $dx\,dt = (x_1 - x_0)(t_1 - t_0)$, we obtain the 1D differential conservation law:

$$u_t + f_x = 0.$$

For $f_x = (H(u))_x$, [i.e., assuming f is a function of u given by $H(u)$] a *weak solution* of the above equation is defined as $u(x,t)$ that satisfies [189]

$$\frac{d}{dt}\int_{x_0}^{x_1} u(x,t)dx = H(u(x_0,t)) - H(u(x_1,t)).$$

Weak solutions are useful in handling nonsmooth data. Observe further that u need not be differentiable to satisfy the above form, and weak solutions are not unique. Thus, we are left with the problem of selecting a special "physically correct" weak solution.

The Hamilton–Jacobi (HJ) equation in $\mathbb{R}^d$ has the form

$$\phi_t + H(\phi_{x_1}, ..., \phi_{x_d}) = 0, \qquad \phi(x,0) = \phi_0(x).$$

Such equations appear in many applications. As pointed out in [187, 188, 159], there is a close relation between HJ equations and hyperbolic conservation laws that in $\mathbb{R}^d$ take the form

$$u_t + \sum_{i=1}^d \frac{df_i(u)}{dx_i} = 0, \qquad u(x,0) = u_0(x).$$

Actually, for the one-dimensional case ($d = 1$), the HJ equation is equivalent to the conservation law for $u = \phi_x$. This equivalence disappears when considering more than one dimension: $H(\cdot)$ is often a nonlinear function of its arguments ϕ_{x_i} and obviously does not have to be separable, so that we can no longer use the integration relation between ϕ and u. However, numerical methodologies that were successfully used for solving hyperbolic conservation laws are still useful for HJ equations.

5.3   Entropy Condition and Vanishing Viscosity

In general, the weak solution for a conservation law is not unique and an additional condition is needed to select the *physically correct* or *vanishing viscosity* solution. This additional condition is referred to as the *entropy condition*.

Consider the "viscous" conservation law:

$$u_t + (H(u))_x = \epsilon\, u_{xx}.$$

The effect of the viscosity $\epsilon\, u_{xx}$ is to smear (or diffuse) the discontinuities, thereby ensuring a unique smooth solution. Introducing the viscosity term turns the equation from a hyperbolic into a parabolic type, for which there always exists a unique smooth solution for $t > 0$. The limit of this solution as $\epsilon \to 0$ is known as the "vanishing viscosity" solution. The entropy condition selects the weak solution of the conservation law

$$u_t + (H(u))_x = 0, \qquad u(x,0) = u_0(x),$$

which is the vanishing viscosity solution for u_0. Therefore, the vanishing viscosity solution is sometimes referred to as the entropy solution.

Satisfying the entropy condition guarantees meaningful and unique weak solutions. Moreover, there is a close duality between the entropy condition and the Eulerian formulation to curve evolution.

5.4 Numerical Methodologies

We have seen that the curve evolution may be presented as a Hamilton–Jacobi equation; see Eq. (4.1). In one dimension, the HJ equation coincides with hyperbolic conservation laws. This close relation can be used to construct numerical schemes for our problems. Similarly to the continuous case, a finite difference method is in *conservation form* if it can be written in the form

$$\frac{u_j^{n+1} - u_j^n}{\Delta t} = -\frac{(g_{j+1/2}^n - g_{j-1/2}^n)}{\Delta x},$$

where $g_{j+1/2} = g(u_{j-p+1}, \dots, u_{j+q+1})$ is called a *numerical flux*, is Lipschitz and *consistent* (satisfies the consistency requirement)

$$g(u, \dots, u) = H(u),$$

that is, setting all the $p + q$ variables of the numerical flux function to u, the numerical flux becomes identical to the continuous flux.

Theorem 3 *Suppose that the solution $u(x, n\Delta t)$ of a finite difference method in conservation form converges to some function $v(x, t)$ as Δx and Δt approach zero. Then $v(x, t)$ is a weak solution of the continuous equation.*

The proof may be found in [200, page 286].

A numerical scheme is *monotone* if the function $F(u_{j-p}^n, \dots, u_{j+q+1}^n)$ that defines the scheme

$$u_j^{n+1} = F(u_{j-p}^n, \dots, u_{j+q+1}^n),$$

or equivalently (for a conservation form):

$$u_j^{n+1} = F(u_{j-p}^n, \dots, u_{j+q+1}^n) = u_j^n - \frac{\Delta t}{\Delta x}(g_{j+1/2}^n - g_{j-1/2}^n),$$

is a nondecreasing function of all its $(p + q + 1)$ arguments.

Theorem 4 [Kuznetsov] [125][1] *A consistent, monotone finite difference method u_j^n that has a conservation form converges to the unique entropy satisfying weak solution of $u_t - (H(u))_x = 0$.*

---

[1]See also [86].

Satisfying the entropy condition is indeed a desired quality, however, these schemes are limited by the following theorem:

Theorem 5 *A monotone finite difference method in conservation form is first-order accurate.*

For proof, see [200, page 299].

Getting higher-order accuracy for such equations by relaxing the monotonicity demand may be found in [161, 159]. One idea leads to the essentially nonoscillating (ENO) schemes, in which an adaptive stencil is used between the discontinuities. Thereby, piecewise smooth data may be handled with high accuracy.

The relation between the Hamilton–Jacobi equations and the conservation laws may be used to design first-order finite difference methods for the HJ equations [161]. Let us explore the relation between $\phi(x,t)$ (the solution of an HJ equation) and $u(x,t)$ (the solution of the corresponding differential conservation law that describes the change of $u =$ "the slope of ϕ"). For the one-dimensional case, it is given by integration, that is, $\phi(x,t) = \int_{-\infty}^{x} u(\tilde{x},t)d\tilde{x}$. Thus by integrating over the monotone numerical scheme (and shifting from $j+1/2$ to j) we arrive at

$$\Phi_j^{n+1} = \Phi_j^n - \Delta t\, g(D_-\Phi_{j-p+1}^n, ..., D_+\Phi_{j+q}^n).$$

Definition 1 *An upwind finite difference scheme is defined so that*

$$g_{j+1/2} = \begin{cases} H(u_j), & H' > 0, \\ H(u_{j+1}), & H' < 0. \end{cases}$$

In some cases, an upwind numerical flux in a conservation form results in a monotone method, though, monotonicity does not necessarily hold for all upwind conservative schemes.

The upwind monotone HJ scheme for the special case where

$$H(u) = h(u^2),$$

with $h'(u) < 0$, was introduced in [161]:

$$g_{HJ}(u_j^n, u_{j+1}^n) = h((\min(u_j^n,0))^2 + (\max(u_{j+1}^n,0))^2).$$

This scheme can be easily generalized to more than one dimension.

Motivated by the theory of mathematical morphology [21], it can be shown that the following scheme has the same qualities (being upwind and monotone) as the HJ scheme under the same restrictions ($h'(u) < 0$):

$$g_M(u_j^n, u_{j+1}^n) = h((\max(-u_j^n, u_{j+1}^n, 0))^2) = h((\min(u_j^n, -u_{j+1}^n, 0))^2).$$

This is also known as a "Godunov scheme" for this case (see, e.g., [161]). The difference between the g_{HJ} and the g_M is that at points where u

changes form negative to positive magnitude, g_M selects the maximum between $(u_j)^2$ and $(u_{j+1})^2$, while g_{HJ} selects $(u_j)^2 + (u_{j+1})^2$. The g_M numerical flux produces better results in some cases.

Having the numerical flux, or numerical Hamiltonian in the HJ context, we can write the numerical approximation of the Hamilton–Jacobi formulation as

$$\Phi_j^{n+1} = \Phi_j^n - \Delta t \, g(D_-\Phi_j^n, D_+\Phi_j^n).$$

As we noted before, in some cases the requirements on the numerical scheme are relaxed to achieve higher-order accuracy as well as handling more complicated flux functions. One useful example for our case is partial derivatives that are approximated by *slope limiters*. The idea is to keep the total variations of the evolving data under control, leading to the TVD (total variation-diminishing) methods [133]. By selecting the smallest slope between the forward and backward derivatives, the estimated slope of the data is always limited by the continuous data. A simple example of a first-order slope limiter is given by the *minmod* operation. Define the minmod selection function as

$$\text{minmod}(a, b) = \left\{ \begin{array}{ll} \text{sign}(a)\min(|a|, |b|) & \text{if } ab > 0, \\ 0 & \text{otherwise.} \end{array} \right.$$

This can be used to approximate ϕ_x by the minmod finite derivative

$$\phi_x|_{x=i\Delta x} \approx \text{minmod}(D_+^x\Phi_i, D_-^x\Phi_i).$$

In [161, 159] it was shown that higher-order accuracy can be easily achieved by using TVD methods for second-order accuracy with solid theory, or using the ENO method for higher-order accuracy (in this case there is not yet a concrete theory for these working schemes). An important implementation issue introduced in [1] is the fact that performing computations only in a narrow band around the propagating front can reduce the computation effort. In this case, computations are performed in a narrow band that is dynamically swept with the front, while the rest of the grid points in the domain serve only as sign holders (see also [43]).

5.5 The CFL Condition

One of the earliest observations in the field of finite difference schemes was made by Courant, Friedrichs, and Lewy in [51, 52]. They observed that a necessary stability condition for a numerical scheme is that the *domain of dependence* of each point in the domain of the numerical scheme should include the domain of dependence of the PDE itself. This condition is necessary, but not necessarily sufficient, for the stability of the scheme. For hyperbolic PDEs the domain of dependence is known to be bounded.

Considering the one-dimensional case, when refining the discretization grid by letting $\Delta x \to 0$ and $\Delta t \to 0$, the ratio $\Delta t / \Delta x$ should be limited. This limit, known as the CFL number or the Courant number, is determined by the maximal possible flow of information. The flow lines of the information obviously depend on the specific initial data and are known as the *characteristics* of the PDE. Collisions of characteristics form "shocks" in the solution and therefore require additional conditions that determine how to handle the propagation of such a shock. A propagating shock in time may thus be defined as a sequence of colliding characteristics where the entropy condition defines the speed of this propagation.

As a simple example consider the one-dimensional conservation law in which the point $(x = \tilde{x}, t = \tilde{t})$ in the PDE domain can be influenced by the data bounded by the triangle $(x_0, 0), (\tilde{x}, \tilde{t}), (x_1, 0)$. This means that any information at the interval (x_0, x_1) of the initial condition u_0 may influence the result at $(\tilde{x}, \tilde{t})$, namely $u(\tilde{x}, \tilde{t})$. Similarly, it may be asserted that the point $(\tilde{x}, \tilde{t})$ is in the *domain of influence* of each point in the interval (x_0, x_1). Therefore, any finite difference approximating the PDE should take this fact into consideration, by limiting the ratio $\Delta t / \Delta x$. Taking this to a limit, for $u_t + (H(u))_x = 0$, the CFL restriction for a three-point scheme can be shown to be

$$1 \geq \frac{\Delta t}{\Delta x} |H'|,$$

and in our case, where we have actually integrated a three-point of Δx scheme of a conservation law into a three-point HJ equation we arrive at the same CFL restriction.

5.6   One-Dimensional Example of a Differential Conservation Law

Let us give a simple example of a one-dimensional differential conservation law where $f(u) = u$, so that

$$u_t = -u_x,$$

given $u(x, t = 0)$. This is a simple case, since the "information" travels along the spatial-temporal lines given by $dx/dt = 1$, which means that the initial data shifts right in time such that

$$u(x, t) = u(x - t, 0);$$

see Figure 5.2. The straight lines in this case are the characteristics. The CFL necessary stability restriction states that the numerical domain of dependence contains the analytic one, that is,

$$\frac{\Delta t}{\Delta x} < 1,$$

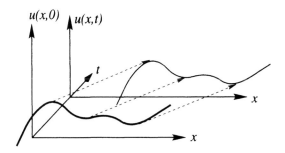

Figure 5.2: The initial data are shifted to the right in time, following $u_t = u_x$.

for a central- or backward-based numerical scheme. See Figure 5.3. The left frame shows a scheme that satisfies the CFL condition for spatial backward temporal forward numerical scheme

$$\frac{U_i^{n+1} - U_i^n}{\Delta t} = -\frac{U_i^n - U_{i-1}^n}{\Delta x},$$

where $U_i^n \approx u(i\Delta x, n\Delta t)$. In the right frame, where the CFL restriction is not satisfied, there is no hope to converge as the numerical grid is refined.

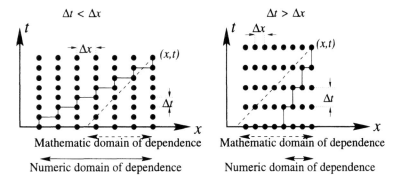

Figure 5.3: The CFL condition: The numerical domain of dependence should include the analytic one, as satisfied on the left and violated on the right.

Now, let us explore the more complicated case

$$u_t = -a(x)u_x,$$

where $a(x)$ is a known function, and say

$$a(x) = \begin{cases} 1, & x < 1, \\ -1, & x > 1. \end{cases}$$

In this case the characteristics slopes are given by

$$\frac{dt}{dx} = \begin{cases} 1, & x < 1 \\ -1, & x > 1. \end{cases}$$

See Figure 5.4.

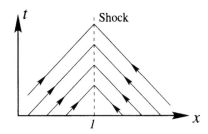

Figure 5.4: The characteristics collide at $x = 1$, forming a shock at a fixed location.

An "upwind" derivative means that as we take the numerical derivative we need to consider the flow of information. For example, the scheme

$$\frac{U_i^{n+1} - U_i^n}{\Delta t} = -\left(\max(a_i, 0) D_x^- U_i^n + \min(a_i, 0) D_x^+ U_i^n \right)$$

is an upwind scheme that converges subject to the CFL condition $\Delta t / \Delta x < a$. This scheme can be alternatively written as

$$U_i^{n+1} = U_i^n - \frac{\Delta t}{\Delta x} \left(\frac{a_i}{2} (U_{i+1}^n - U_{i-1}^n) - \frac{|a_i|}{2} (U_{i+1}^n - 2U_i^n + U_{i-1}^n) \right).$$

The second term can be associated with the viscosity term as $\Delta x |a_i| u_{xx} = \epsilon |a| u_{xx}$. The numerical derivatives are taken w.r.t. the direction from which the information is flowing. In other words, the "data" are carried by the "wind" and thus the term "upwind."

5.7   Two-Dimensional Example of the CFL Condition

As pointed out, the g_{HJ} and the g_M numerical flows may be easily generalized to several dimensions. The generalization is straightforward and for the specific case of $H(u, v) = f(u^2, v^2)$ we get the following form:

$$g_M(u_i^n, u_{i+1}^n, v_j^n, v_{j+1}^n) = h((\max(-u_i^n, u_{i+1}^n, 0))^2, (\max(-v_j^n, v_{j+1}^n, 0))^2).$$

This yields an upwind monotone scheme with a CFL restriction of

$$1 \geq \left(\frac{\Delta t}{\Delta x} |H_u| + \frac{\Delta t}{\Delta y} |H_v| \right).$$

Consider the simple example of a planar curve propagating with constant velocity along its normal that obeys the following evolution law:

$$C_t = \vec{N}.$$

It is easy to see that since $V_N = 1$ the Eulerian formulation for this case is

$$\phi_t = |\nabla\phi|;$$

thus, $H(u, v) = \sqrt{u^2 + v^2}$. For the simple selection of $\Delta x = \Delta y = 1$, we arrive at the CFL restriction:

$$\Delta t \leq \frac{1}{\sqrt{2}}.$$

The following example presents offsets produced by two schemes, one with $\Delta t < 1/\sqrt{2}$, satisfying the CFL restriction, and another with $\Delta t > 1/\sqrt{2}$, violating the CFL restriction. The Eulerian formulation is implemented by the following numerical approximation:

$$\Phi_{ij}^{n+1} = \Phi_{ij}^{n} + \Delta t \Big(\big(\max(-D_{-}^{x}\Phi_{ij}^{n}, D_{+}^{x}\Phi_{ij}^{n}, 0)\big)^{2}$$
$$+ \big(\max(-D_{-}^{y}\Phi_{ij}^{n}, D_{+}^{y}\Phi_{ij}^{n}, 0)\big)^{2}\Big)^{1/2}.$$

The left image Φ^0 in Figure 5.5 is given as initial condition to the evolution

Figure 5.5: The images of the iterations (every two time steps) Φ^0 to Φ^{14}, left to right, for the scheme with time step $\Delta t = 0.7$ that satisfies the CFL restriction

equation. The evolution of Φ in time for the scheme with $\Delta t = 0.7 < 1/\sqrt{2}$ is presented in the following frames in Figure 5.5. The offset results of the two schemes with $\Delta t = 0.7$ and $\Delta t = 0.8$ are presented in Figure 5.6 on the top and bottom rows, respectively. The gray levels correspond to the height values of Φ_{ij}^{n} on the grid. Histogram equalization is applied to the last evolution step to enhance the fact that violation of the CFL restriction results in perturbations of the Φ function. The right column in Figure 5.6 shows the unstable result at the bottom compared with the stable one at the top. The zero level sets (every two time steps) are drawn as white contours on the original image (left column).

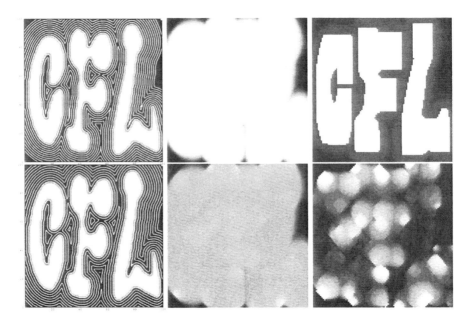

Figure 5.6: Top: $\Delta t = 0.7$. Bottom: $\Delta t = 0.8$, violating the CFL condition. Left: The offsets (zero level sets, of the propagating Φ every two time steps) are shown as white contours on the original image. Middle: The images of Φ at $t = 11.2$, in which the heights Φ_{ij} are presented as gray levels. Right: Φ images at $t = 11.2$ after histogram equalization that stresses the instability effects caused by violating the CFL condition.

5.8   Viscosity Solutions

The terminology of vanishing viscosity was adopted to a more general class of solutions known as viscosity solutions. Based on a simple comparison principle, solutions can be tested for uniqueness and existence, while numerical schemes can be designed to extract the vanishing viscosity solution. It is beyond the scope of this book to review the whole theory of viscosity solution, for which we refer the reader to [54].

Let us consider the following example:

$$u_t(x, t) = F(\nabla u(x, t), x),$$

where F is a continuous function. We say that u is a *viscosity subsolution* at (x_0, t_0) if for all smooth ψ, such that (x_0, t_0) is a local maximum of $u - \psi$, we have

$$\psi_t(x_0, t_0) \leq F(\nabla \psi(x_0, t_0), x).$$

In a similar way, u is a *viscosity supersolution* at (x_0, t_0) if for all smooth ψ, such that (x_0, t_0) is a local minimum of $u - \psi$, we have

$$\psi_t(x_0, t_0) \geq F(\nabla\psi(x_0, t_0), x).$$

Next, u is a viscosity solution at (x_0, t_0) if it is both a viscosity subsolution and a supersolution. Finally, u is a viscosity solution if it satisfies the initial conditions, and it is a viscosity solution for all (x, t).

This comparison principle can be used to design numerical schemes that pick up the vanishing viscosity solution. One example that will be useful in the following chapters is based on [9, 170]. It states that a monotone, stable, and consistent numerical scheme converges to the correct viscosity solution if it satisfies a comparison principle. In this case the subsolution and super-solution are used to define a *lower semicontinuous viscosity supersolution*, denoted as lsc or $\underline{u}$, and an *upper semicontinuous viscosity subsolution*, denoted as usc or $\overline{u}$. The comparison principle leads to a *strong uniqueness* property.

Practically, when we construct a numerical scheme for the equation

$$F(\nabla u, x) = 0,$$

for which there is a viscosity solution, it is sufficient to verify that a consistent and stable approximation of the form

$$g_{ij}(U_{ij}, U_{i+1,j}, U_{i-1,j}, U_{i,j+1}, U_{i,j-1}, i, j) = 0$$

is monotone in the sense that if $U \geq V$ for all i, j, then

$$g_{ij}(w, U_{i+1,j}, U_{i-1,j}, U_{i,j+1}, U_{i,j-1}, i, j) \leq$$
$$g_{ij}(w, V_{i+1,j}, V_{i-1,j}, V_{i,j+1}, V_{i,j-1}, i, j),$$

for all $w \in \mathbb{R}$. We later show how this monotonicity principle can be exploited to construct very efficient sequential schemes for solving eikonal equations.

5.9 Summary

We reviewed the basic terminologies and methodologies in numerical analysis of conservation laws and viscosity solution. Following Osher and Sethian, it was shown how planar curve evolution can be casted into the Eulerian formulation. This implicit formulation for curve evolution has the form of a Hamilton–Jacobi type of equation, for which there is a close relation to conservation laws. This relation was then explored and used to achieve efficient and stable numerical schemes.

5.10   Exercises

1. Define a second-order numerical approximation for u_x at $x = i\Delta x$, based on u_i, u_{i+1}, u_{i+2}.

2. Define a numerical approximation for u_{xx} at $x = i\Delta x$ based on u_i, u_{i+1}, u_{i+2}. What is the order of the truncation error for your approximation?

3. Apply the Osher–Sethian level set method to the constant flow. Specify the numerical approximation scheme you used, and report numerical difficulties.

4. Use Green's theorem to prove the divergence theorem, and derive the differential conservation law.

6

Mathematical Morphology and Distance Maps

Mathematical morphology defines algebraic operations on geometric sets (shapes). The basic operations that correspond to addition and subtraction are the *dilation* δ and the *erosion* ϵ operations that are denoted as $\oplus$ and $\ominus$, respectively. For our purposes a shape is defined as the interior of a closed curve. Exploring the history of mathematical morphology that started with the work of G. Matheron and J. Serra, and the various applications especially in image synthesis and processing, is beyond the scope of this chapter. The interested reader could start with a simple introduction to this field in Dougherty's book [60]. Here we concentrate on some of the basic operations. Motivated by [21, 177], and the related slope transform of P. Maragos [148], and Dorst and Boomgaard [59], we explore the link between curve evolution and continuous-scale morphology. We also refer to [77, 76], for efficient algorithms of the basic morphological operations with a rectangular structuring element.

In order to define operations between shapes, one shape is referred to as a *structuring element*. A structuring element is defined by a shape and a point of reference, an origin, that we sometimes refer to as the "center." A simple example is a circle with an origin at its center.

The dilation of the shape A with the structuring element B, denoted as $\delta_B(A) = A \oplus B$, is the envelope of the shape obtained by placing copies of B in A such that the center is restricted to be in A. The dilation $\delta_B : \mathbb{R}^N \to \mathbb{R}^N$ of the set $A \subset \mathbb{R}^N$ by the structuring element $B \subset \mathbb{R}^N$ is defined as

$$
\begin{aligned}
\delta_B(A) &\equiv A \oplus B \equiv \bigcup_{b \in B} \{(A + b)\} \\
&= \{a + b : a \in A, b \in B\}.
\end{aligned}
$$

Thereby, the dilation of the shape A with the structuring element B, which is a circle of radius r, yields a shape boundary given by an offset of distance r from the boundary of A. See Figure 6.1.

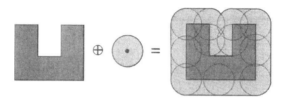

Figure 6.1: Dilation of the shape A with the circle B as a structuring element.

Similarly, erosion $\epsilon_B : \mathbb{R}^N \to \mathbb{R}^N$ of the set $A \subset \mathbb{R}^N$ by the structuring element $B \subset \mathbb{R}^N$ is defined as

$$
\begin{aligned}
\epsilon_B(A) \equiv A \ominus B \ &\equiv \ \bigcap_{b \in B} \bigcup_{a \in A} \{a - b\} \\
&= \ \{x : x + b \in A, \forall b \in B\} \\
&= \ \{x : x + B \subset A\} = \bigcap\{A - b : b \in B\}.
\end{aligned}
$$

Other useful basic operations are *Opening*, defined by sequential application of erosion followed by dilation

$$
A \circ B = \delta_B(\epsilon_B(A)) = (A \ominus B) \oplus B,
$$

which smoothes the shape, and *Closing*, defined as

$$
A \bullet B = \epsilon_B(\delta_B(A)) = (A \oplus B) \ominus B,
$$

which smoothes the background. See Figure 6.2.

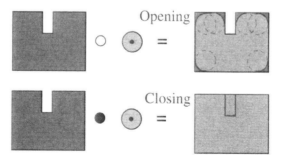

Figure 6.2: Opening and closing of the shape A with the circle B as a structuring element.

6.1 Continuous Morphology by Curve Evolution

Fundamental properties of the dilation are its commutativity, $A \oplus B = B \oplus A$, and associativity

$$(A \oplus B) \oplus D = A \oplus (B \oplus D).$$

Before we apply a convex structuring element recursively on shapes, let us show that

Lemma 6 [Alvarez–Guichard–Lions–Morel] [3] *For any* $t_1, t_2 \geq 0$, *we have*

$$(t_1 + t_2)B = t_1 B \oplus t_2 B$$

(where $tB = \{tx : x \in B\}$, *for* $t \in \mathbb{R}^+$*) if and only if* B *is convex.*

Proof. If B is convex, then for every $t_1, t_2 \geq 0$ we have by convexity that $t_1 B \oplus t_2 B \subseteq (t_1 + t_2)B$ and always that $t_1 B \oplus t_2 B \supseteq (t_1 + t_2)B$. See Figure 6.3.

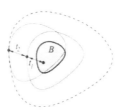

Figure 6.3: If the structuring element B is convex, then $t_1 B \oplus t_2 B \subseteq (t_1 + t_2)B$ and $t_1 B \oplus t_2 B \supseteq (t_1 + t_2)B$.

Now, if $t_1 B \oplus t_2 B \subseteq (t_1 + t_2)B$, then for every two points, $p_1, p_2 \in B$, we can find $p_3 \in B$, such that $(t_1 + t_2)p_3 = t_1 p_1 + t_2 p_2$. It means that

$$p_3 = \frac{t_1}{t_1 + t_2}p_1 + \frac{t_2}{t_1 + t_2}p_2,$$

which is a point along the line connecting p_1 and p_2, and therefore should also be in B. This should hold for every selection of t_1 and t_2, which is the definition of convexity. Thus, B must be convex. See Figure 6.4. ■

We can now dilate shapes by recursive application of convex structuring elements using the following property:

Lemma 7 *For a convex B we have*

$$\delta_{(t_1 + t_2)B}(X) = \delta_{t_1 B}(\delta_{t_2 B}(X)),$$

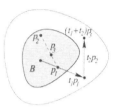

Figure 6.4: A recursive definition of a structuring element holds if and only if it is convex.

where for $t \in \mathbb{R}^+$, $tB = \{tx : x \in B\}$.

Proof. By convexity (Lemma 6) and the associativity, we have that

$$
\begin{aligned}
\delta_{(t_1+t_2)B}(X) &= \bigcup_{b \in (t_1+t_2)B} \bigcup_{x \in X} \{x + b\} \\
&= X \oplus (t_1 + t_2)B \\
&= X \oplus (t_1 B \oplus t_2 B) \\
&= (X \oplus t_1 B) \oplus t_2 B \\
&= \delta_{t_1 B}(\delta_{t_2 B}(X)).
\end{aligned}
$$
∎

We can now use the associativity property and the above lemma to decompose convex structuring elements into infinitesimal self-similar copies that are applied sequentially. That is, any convex B is the result of n sequential dilations of a point (note the analogy to impulse response in classical linear signal processing) with the scaled structuring element $\tilde{B} = \frac{1}{n}B = \{b : nb \in B\}$. It is easy to prove this claim using the above lemma. Note also that for a convex B we readily have that

$$
\begin{aligned}
X \oplus \overbrace{\tilde{B} \oplus \tilde{B} \oplus \cdots \oplus \tilde{B}}^{n} &= \delta_{\tilde{B}}(\delta_{\tilde{B}}(\dots(\delta_{\tilde{B}}(X))\dots)) \\
&= \delta_{(\frac{1}{n}+\cdots+\frac{1}{n})B}(X) \\
&= X \oplus B = \delta_B(X).
\end{aligned}
$$

Recalling curve evolution properties, we are ready to define continuous-scale morphology by curve evolution.

6.2 Continuous-Scale Morphology

Let us zoom in to the local behavior along the boundary of a given shape while applying a dilation operation with an infinitesimal convex structuring element. We have seen that successive application of the structuring element amounts to the dilation by its scaled version. Figure 6.5 shows the normal velocity that results from the dilation of a shape A with the structuring element B. The boundary of A is given by the curve C, while the structuring element B is given in polar coordinates with respect to its center by $r(\theta)$.

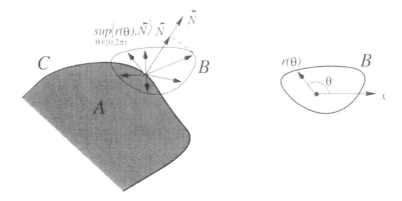

Figure 6.5: The normal velocity for dilation of A with a convex structuring element B.

We can consider each point along the curve as a spreading process of many points that propagate toward the structuring element's boundary. We have learned earlier by the Gage and Epstein theorem that the geometry of an evolving curve is determined only by the normal component of its velocity. Now, consider all possible velocities that are generated by the dilation operation at a given point along the shape boundary, and consider only their corresponding normal components. In order to find the envelope, we need to consider only the maximal normal component.

Formally, we have the continuous-scale morphological dilation of a shape A defined by its boundary C given by the boundary curve evolution equation

$$\frac{dC}{dt} = \sup_{\theta \in [0,2\pi]} \langle r(\theta), \vec{N} \rangle \vec{N},$$

with the boundary of the original shape as initial condition $C(0) = C$. Here, $t \in [0, T]$ represents the scale of the structuring element, that is, $\partial B(t) = r(\theta)t$.

The simplest example is a continuous dilation with a circular structuring element $r(\theta) = 1$, in which case, the normal component is given by $\sup_{\theta \in [0,2\pi]} \langle r(\theta), \vec{N} \rangle = 1$, and the evolution rule is the constant flow

$$\frac{dC}{dt} = \vec{N}.$$

Indeed, the dilation with a circular structuring element results in offset curves to the boundary of the original shape. Another way to look at the boundaries formed by morphological dilation with continuous scale is as the level sets of a distance map, in which the structuring element defines the unit sphere.

6.3 Distance Maps

In many applications it is important to have a measure of distance from a set of points or objects. A distance map assigns a value that corresponds to the shortest distance from a given set of source points to each point in the domain. Both morphological dilation and erosion operations with scaled versions of convex structuring elements can be interpreted as level sets of the distance from the shape's boundaries. The "unit circle" in this case is defined by the structuring element. That is, the structuring element defines the local metric by $ds^2 = r(\theta)^2 \sqrt{dx^2 + dy^2}$. As a simple example, dilation with a circular structuring element results in an offset of the boundary, given by the Euclidean distance defined by the arclength $ds^2 = dx^2 + dy^2$.

Assume we have a given set of points P, that might just as well represent a curve. Then, the distance map $T_P : \mathbb{R}^2 \to \mathbb{R}_+$ assigns a value to each point in the domain such that

$$T_P(q) = \inf_{p \in P} d(p, q),$$

where $d(p, q) = \|p - q\|_{L_2}$ is the Euclidean distance between the two points p and q.

Let us limit our world and allow operations only at a set of "pixels" described by a rectangular grid sampling the continuous plane. Now assume we have one given source point $p \in P$ located at one of the pixel points. The distance map computation involves visiting each grid point q, and computing $d(p, q) = \|p - q\|_{L_2} = \sqrt{(p_1 - q_1)^2 + (p_2 - q_2)^2}$. If we have N points in our grid, the distance map computation complexity is $O(N)$ without error. Now let us add more points to our source points set P, say half of the pixels in the domain. In principle, we need to compute the distance from each point to all the source points in P and assign the minimal distance result. However, we can use the fact that the distance map "characteristics" are straight lines and come up with an $O(N)$ procedure known as the Danielsson distance map algorithm [55].

The basic idea is to use a four-scan method to propagate the distance information in four sections. The scans are left-to-right top-to-bottom, right-to-left top-to-bottom, left-to-right bottom-to-top, and right-to-left bottom-to-top. Each point stores two integer numbers that indicate its relative location to the closest source pixel.

All source pixels are set to $(0,0)$, while the nonsource pixels are initially set to (∞,∞). In each scan every pixel compares its source indicator with its neighbors and adopts their source point if appropriate. This way each grid point "tells" its neighbors where the closest source is.

Since we have a four-scan method for computing the distance, the result is an $O(N)$ algorithm. Yet, small errors may occur along the singularities of the distance map. These points are known as medial axis, skeleton, Voronoi diagram edges, or sometimes referred to as the edges of the "pseudo Dirichlet tessellation" of the source points.

Practically, it can happen that a grid point has a $(3,4)$ relative offset from one source point, and $(0,5)$ from another, such that the distance is the same to both. Yet the order of the scanning prefers one source over the other and blocks the correct propagation of information to one grid point along the medial axis; see Figure 6.6.

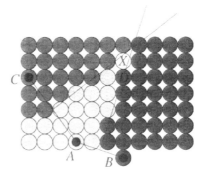

Figure 6.6: The points A, B, and C are source points, and the pixels are colored according to their source point association. Since the point D shares A and B as closest points, it can be associated with B that blocks the correct update to the point X. The value of X would be 6 instead of $\sqrt{35} = 5.916$.

Let us further refine the problem. Assume that the distance we measure is no longer from a set of feature grid points, but the distance now refers to subpixel features with subpixel resolution. That is, at some given grid points we have a distance measure from object boundaries that exist between the pixels. We would like to use this information and find the distance to these subpixel boundaries at the rest of the domain. Observe that the initial conditions introduce an error of $O(\Delta x)$ on the boundary location.

In Chapter 7 we explore the differential characteristics of distance maps, namely the fact that $|\nabla T| = 1$ almost everywhere, and show that the numerical approximation

$$|\nabla T|^2 \approx \left(\max(-D_+^x T_{ij}, D_-^x T_{ij}, 0)\right)^2 + \left(\max(-D_+^y T_{ij}, D_-^y T_{ij}, 0)\right)^2$$

selects the desired solution. It can therefore be used as part of a linear alternating scanning directions algorithm, similar to Danielsson's, for the distance computation in subpixel accuracy. At this point we assume we have an efficient distance computation procedure, either by linear alternating scanning or a fast marching method-based approach that we explore in Chapter 7. We next show how to apply these procedures to efficiently locate the symmetry axis of objects embedded in gray-level images.

6.4   Skeletons

Our visual perception relies on symmetry as an important cue for image representation and understanding. In [16, 17], motivated by visual perception, Blum describes the medial axis transform for shape description and analysis. Early mathematical properties and procedures for the computation of the symmetry axis, also known as *skeleton*, can be found in [150, 151]. The topic is still of great interest to the computer vision community.

Motivated by [115], we review a level set approach for the computation of the medial axis. Unlike most of the existing methods, this approach relies on an initial segmentation of the shape's boundary. This initial segmentation provides a global numerical support for an accurate, stable, and efficient computation of the skeleton.

Before we introduce the method let us follow Montanari [150] and give four equivalent definitions for a skeleton of a shape in the real plane. The first is the prairie-fire model: Skeleton points are defined by the locations at which the propagating wavefront, initiated at the shape's boundary, intersects itself. The boundary propagates with a constant velocity along its normal direction; the skeleton is the set of shocks or curvature singularities that appear as two normals collide.

The skeleton is also described as the projection of the distance map ridges, as the centers of maximal disks, and as the set of points that do not belong to lines connecting other points to their corresponding closest boundary point. See Figure 6.7.

Using the above definitions, we see that each skeleton point has the same shortest distance from at least two boundary points. If we segment the boundary to many small segments, measure the distance from each of these segments, then the skeleton points belong to the zero sets of the distance maps differences.

We implicitly assume that we know how to segment the boundary so that the zero sets include the skeleton. One way to satisfy this assump-

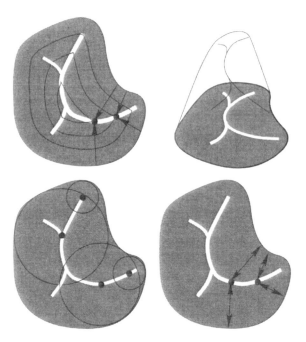

Figure 6.7: Four ways to describe a skeleton. Top left: Symmetry axis as shocks along the propagating boundary formed at the locations of colliding normals. Top right: The skeleton is the projection of the ridges of the distance map. The third dimension represents the distance. Bottom left: The skeleton is the set of centers of maximal disks. Bottom right: The skeleton is the set of points that do not belong to any straight line connecting another point to its closest boundary point.

tion is to take the segmentation to the limit, consider the set of pixels or points that approximate the boundary, and construct a Voronoi diagram skeletonization [130, 20, 154]. In order to extract the actual skeleton, these methods involve extensive pruning of the symmetry sets of neighboring boundary pixels, which we would like to avoid.

A skeleton corresponds to at least two boundary points that touch a maximal disk. We say that these boundary points are "generating" the skeleton point. A useful relation between the skeleton and the boundary states that

Theorem 6 *A boundary segment that connects two points that generate the same skeleton point contains at least one internal point with a positive curvature maximum.*

Proof. The maximal disk that corresponds to the skeleton point is tangent to the boundary at the endpoints of the boundary segment. The two tangent points also segment the disk into two arcs. Consider the curve defined by the boundary segment and the complementary arc of the disk. Next, we apply a special version of the four-vertex theorem, see [115], by which a closed regular curve contains at least two distinct points at which the curvature is both positive and locally maximal. The circular arc of our new curve may account for one such point. The other is located along the segment of the boundary. See Figure 6.8. ∎

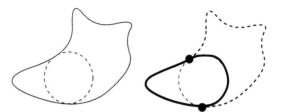

Figure 6.8: Left: The original boundary curve and a maximal disk. Right: The new curve defined by the circular arc and the boundary segment, with at least one positive curvature maximum.

The above theorem provides the basis for the necessary boundary segmentation. Namely, if we segment the boundary at points of positive maximal curvature, generating points that correspond to the same skeleton point belong to two different segments.

We are now ready to introduce our skeletonization algorithm.

1. Compute the curvature along the boundary curve C and segment the boundary at points of maximal positive curvature. Assume there are N such segments, defined as C_i, where $i \in \{0, \dots, N-1\}$.

2. Compute the distance map T_i for each boundary segment defined as

$$T_i(x, y) = \inf_{p \in C_i} \{d(p, (x, y))\} .$$

3. Find a preliminary skeleton as the zero level set of all distance map differences.

$$\hat{S} = \{(x, y) : \forall i, j; i \neq j, (T_i(x, y) - T_j(x, y) = 0)$$
$$\cap (\bigcap_{k \neq i,j} \{T_i(x, y) < T_k(x, y)\})\},$$

where $i, j, k \in \{0, 1, \ldots, N - 1\}$. $\hat{S}$ is the Voronoi diagram of the boundary segments.

4. Delete all external background points and the skeleton branches starting at the segment point i with a distance less than $1/\kappa_i$ from the segment endpoint, where κ_i is the curvature at the end of each segment. This pruning process starts with the points of highest curvature.

This algorithm works well for all planar shapes. Yet, care must be taken when dealing with holes. The main reason is that the boundary of a hole does not have to satisfy Theorem 6. A general fix for this problem is to add a segmentation point at a point with turn angle of 2π along the boundary from a local positive curvature maximum point. That is, integrate the curvature along the arclength and initiate a new segment when the result is more than 2π. This guarantees a correct and minimal boundary segmentation necessary for the skeleton construction. See Figure 6.9.

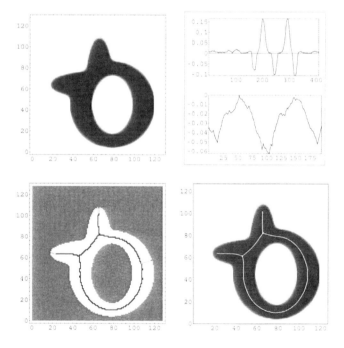

Figure 6.9: Top left: Original image. Top right: The curvature of the outer and inner boundaries as a function of their arclength. Bottom left: Pixel level skeleton with branch pruning. Bottom right: Subpixel skeleton.

All calculations can be performed at a pixel level as a numerical approximation for the continuous case. That is, the curvature of the threshold level set can be approximated at the pixel points as

$$\kappa = \mathrm{div}\left(\frac{\nabla I}{|\nabla I|}\right),$$

where I is a gray-level image of the shape. The zero set of the distance maps differences yields a global numerical support for the skeleton location. Finally, the skeleton itself can be linearly interpolated between the pixel grid points.

6.5 Exercises

1. Find the curve evolution rules that correspond to continuous morphological dilation with a circular structuring element, a square, a diamond, and an ellipse with given properties (orientation and aspect ratio).

2. What is the level set formulation for each of the above evolution equations?

3. Implement the above continuous-scale morphology via the level set framework and apply it to a nontrivial shape. Implement opening and closing, and report numerical difficulties.

4. Research project: Define and apply mathematical morphology to shapes on smooth manifolds.

5. Use Danielsson's Euclidean distance map to compute the distance map of a binary shape. Compute the skeleton based on the set of singularities of the distance map and the connectivity of the skeleton.

6. Apply the zero level set approach by distance differences for skeletonization. Use the fast marching method to compute the distances from the boundary segments of a shape given as a threshold of a gray-level image.

7. Find the skeleton that is invariant to the shear $\begin{pmatrix} 1 & a \\ 0 & 1 \end{pmatrix}$ transform. If it is not connected, modify it to be one.

Fast Marching Methods

The fast marching method was introduced by Sethian [190, 191, 192] as a computationally efficient solution to *eikonal equations* on flat domains. A related method was presented by Tsitsiklis in [205]. The fast marching method was extended to triangulated surfaces by Kimmel and Sethian in [112]. The extended method solves eikonal equations on flat rectangular or curved triangulated domains in $O(M)$ steps, where M is the number of vertices. In other words, the computational complexity of finding the solution is optimal. Here, we will present an $O(M \log M)$ method that is simple to implement. Using this technique, one can efficiently compute distances on curved manifolds with local weights. In this chapter we start with a simple example of distance computation in 1D that simplifies the notion of viscosity solutions of eikonal equations. Next, the fast marching method is applied to path planning of a robot navigating in nontrivial configuration space with a small number of degrees of freedom. Finally, we explore the power of efficient distance computation on curved domains, and present applications like minimal geodesic computation on triangulated surfaces, geodesic Voronoi diagrams, and curve offset calculations on weighted surfaces.

In some problems we encounter an eikonal equation, which is a differential formulation of a wave propagation equation, that in 2D reads

$$|\nabla T(x, y)| = \mathcal{F}(x, y).$$

Here $\mathcal{F} : \mathbb{R}^2 \to \mathbb{R}^+$ is given, and we are looking for $T : \mathbb{R}^2 \to \mathbb{R}^+$, with given $T(x_0, y_0) = 0$ as boundary conditions. Other boundary conditions may be T values given at any subset of $\mathbb{R}^2$, in which case we need additional assumptions to the way we expect the solution to behave. We mentioned in Chapter 5 that the following approximation of the gradient magnitude results in a monotone scheme that selects the viscosity solution:

$$|\nabla T(x,y)|^2 \approx \left(\max\{-D^x_+T_{ij}, D^x_-T_{ij}, 0\}\right)^2$$
$$+ \left(\max\{-D^y_+T_{ij}, D^y_-T_{ij}, 0\}\right)^2.$$

Let us first give some intuition behind this approximation in 1D, and link between weighted distance and eikonal equations, that in 1D reads

$$|T_x(x)| = \mathcal{F}(x),$$

where $\mathcal{F} : \mathbb{R} \to \mathbb{R}^+$ is a given positive function, and say $T(x_0) = 0$.

7.1   The One-Dimensional Eikonal Equation

In order to gain some motivation, let us consider the following simple example in one dimension. We would like to compute the distance from a single point x_0 on the line $\hat{x}$. Let us use the $\hat{y}$-axis as a distance indicator, and search for the distance function $T(x)$, where we already know that $T(x_0) = 0$. The distance at the point x is given by $T(x) = |x - x_0|$ as shown in Figure 7.1, left. Taking the magnitude of the first derivative of the distance function T, we get $|\partial_x T| = 1$, except at the point x_0. Let us add a second point x_1, where $T(x_1) = 0$. The distance function from the two points is now given by $T(x) = \min\{|x - x_0|, |x - x_1|\}$; see the solid line in Figure 7.1, right. Again $|\partial_x T| = 1$ characterizes the distance almost everywhere, except at x_0, x_1 and the point $(x_0 + x_1)/2$.

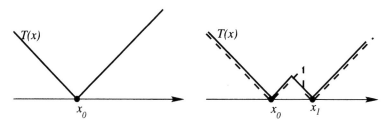

Figure 7.1: The distance from a point at x_0 is given by $T(x) = |x - x_0|$, while the distance from two points is given by $T(x) = \min\{|x-x_0|, |x-x_1|\}$.

Let us formulate the distance map by its derivative equation $|\partial_x T| = 1$, which is a one-dimensional eikonal equation, with boundary conditions $T(x_0) = T(x_1) = 0$. We would like to construct numerical schemes that compute T that satisfies the eikonal equation "the best." For example, the dashed line in Figure 7.1, right, satisfies $|\partial_x T| = 1$ except for three points, yet the discontinuity introduced by the dashed line is less smooth than the solid line. The smoothest solution is referred to as the viscosity solution, and the question is how to numerically select the viscosity solution among all other "weak" solutions.

Now, say we restrict the computation to points on a uniform discrete grid, $T_i = T(ih)$, where h is the grid spacing. Let us test some numerical approximations for the one-dimensional eikonal equation. Consider first the backward approximation

$$|\partial_x T| \approx \left| \frac{T_i - T_{i-1}}{h} \right|.$$

Then, a possible solution for $|T_i - T_{i-1}| = h$ is shown in Figure 7.2, left.

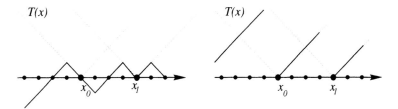

Figure 7.2: Left: A possible numerical solution for $|T_i - T_{i-1}| = h$. Right: A possible numerical solution for $|\max\{T_i - T_{i-1}, 0\}| = h$.

We see that a better approximation should be applied. Next, in order to exclude decreasing updates we test the approximation

$$|\partial_x T| \approx \left| \max\left\{ \frac{T_i - T_{i-1}}{h}, 0 \right\} \right|.$$

A possible solution is shown in Figure 7.2, right. This approximation allows the point i to change its value only to a value greater than its left neighbor's, T_{i-1}. Let us add symmetry, and use the approximation

$$|\partial_x T| \approx \left| \max\left\{ \frac{T_i - T_{i-1}}{h}, \frac{T_i - T_{i+1}}{h}, 0 \right\} \right|.$$

Now, the update of a point i allows the point to get a higher value than the value of the neighboring point that is used for its update. This approximation may be equivalently written as

$$|\partial_x T| \approx \left| \max\left\{ \frac{T_i - \min\{T_{i-1}, T_{i+1}\}}{h}, 0 \right\} \right|.$$

This numerical approximation is known as "upwind," since it senses the direction of information flow, from which the "wind blows," and selects the derivative with respect to the smaller neighboring value. It selects the viscosity solution as desired and displayed in Figure 7.3.

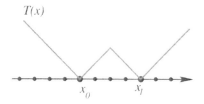

Figure 7.3: The numerical viscosity solution for
$|\max\{T_i - \min\{T_{i-1}, T_{i+1}\}, 0\}| = h.$

A general scheme that solves this last approximation is

$\forall i, T_i \leftarrow \infty$
$k = x_0/\Delta x$
$l = x_1/\Delta x$
$T(k) = T(l) = 0$
repeat until convergence, $\forall i, i \neq k, i \neq l$
$\qquad t = h + \min\{T_{i-1}, T_{i+1}\}$
$\qquad$ if $T_i > t$ then $T_i = t.$

The next question is what order of updates should be applied such that
the solution is obtained with minimal update steps. If we scan the line
successively from left to right, we need as many scans as the number of
grid points. It guarantees that in the worst-case scenario, where $T(x) = 0$
at the rightmost point, the distance will propagate its way to the leftmost
point.

A better scanning strategy would be to apply first a left to right followed
by a right to left update scan. Two scans are sufficient to integrate the
distance from the source points and compute the solution in this case.
Actually, the one-dimensional eikonal equation suggests two possible slopes
at each point, while the ordered integration selects the proper one. An
extension of this alternating scanning direction method is used to compute
Euclidean distances in higher dimensions; see, for example, [55]. Another
efficient order of updates, which is somewhat more natural in the case of
distances, follows the wave front. It starts from the initial points x_0 and x_1,
and updates their neighboring points. First, set the T values of all points
to infinity, and update the value at the points with a given initial value.
Next, the updated point that touches points that were not updated and
has the smallest value is selected, and its neighbor is updated. This last
step repeats until there are no more points to update. The fast marching
method is an extension of this procedure.

Let us explore some aspects of distance computation on weighted domains. In order to compute the distance between two points we need to define a measure of length. A definition of an arclength allows us to measure distance by integrating the arclength along a curve connecting two points. The distance between the points corresponds to the shortest length among all curves connecting them.

Length is now influenced by a local weight $\mathcal{F}(x) : \mathbb{R} \to \mathbb{R}^+$ at each point, where the arclength is defined by $ds = \mathcal{F}(x)dx$. We search for the distance function, characterized by the eikonal equation $|\partial_s T(x)| = 1$. By the chain rule, $\partial_s T = \partial_x T \partial_s x = (1/\mathcal{F}(x))\partial_x T$, such that the eikonal equation may be written as

$$|\partial_x T(x)| = \mathcal{F}(x);$$

see Figure 7.4. The above numerical approximation selects again the viscosity solution in this case.

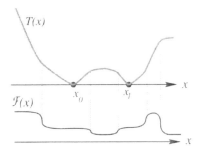

Figure 7.4: The single numerical solution for $|\max\{T_i - \min\{T_{i-1}, T_{i+1}\}, 0\}| = h\mathcal{F}_i$.

7.2 Fast Marching on Two-Dimensional Rectangular Grids

Given a two-dimensional weighted flat domain, or in other words an isotropic non-homogeneous domain, the distance is defined via the arclength. For example, the arclength may be written as a function of the x and y Cartesian coordinates of the planar domain

$$ds^2 = \mathcal{F}^2(x, y)(dx^2 + dy^2),$$

where $\mathcal{F}(x, y) : \mathbb{R}^2 \to \mathbb{R}^+$ is a function that defines a weight for each point in the domain.

The distance map $T(x, y)$ from a given point p_0 assigns a scalar value to each point in the domain that corresponds to its distance from p_0. It is easy to show (see, e.g., [11]) that like in the 1D case, the gradient magnitude of the distance map is proportional to the weight function at each point

$$|\nabla T(x, y)| \;=\; \mathcal{F}(x, y), \tag{7.1}$$

where $|\nabla T| \equiv \sqrt{T_x^2 + T_y^2}$. This is an eikonal equation in 2D. The viscosity solution to the eikonal equation coupled with the boundary condition $T(p_0) = 0$ results in the desired distance map.

Our first goal is to solve the eikonal equation. Again, we would like to construct a numerical approximation to the gradient magnitude that selects an appropriate "weak solution." The use of upwind schemes in hyperbolic equations is well known; see, for example, the Godunov paper from 1959 [78]; as well as for Hamilton–Jacobi equations (see, e.g., [170, 21]). Consider the following upwind approximation to the gradient, given by

$$\left(\left(\max\{D_{ij}^{-x}T, -D_{ij}^{+x}T, 0\} \right)^2 + \left(\max\{D_{ij}^{-y}T, -D_{ij}^{+y}T, 0\} \right)^2 \right)^{1/2} = \mathcal{F}_{ij}, \tag{7.2}$$

where $D_{ij}^{-x}T \equiv (T_{ij} - T_{i-1,j})/h$ is the standard backward derivative approximation, and $T_{ij} \equiv T(i\Delta x, j\Delta y)$.

We use the following relation

$$\max\{D_i^- T, -D_i^+ T, 0\} = \max\{T_i - \min\{T_{i-1}, T_{i+1}\}, 0\},$$

where we assume $h = \Delta x = 1$. Now, the approximation (7.2) can be written as

$$(\max\{T_{ij} - \min\{T_{i-1,j}, T_{i+1,j}\}, 0\})^2$$
$$+ (\max\{T_{ij} - \min\{T_{i,j-1}, T_{i,j+1}\}, 0\})^2 = \mathcal{F}_{ij}^2,$$

where, w.l.o.g., we assume $h = \Delta x = \Delta y = 1$, and the update step is given by the T_{ij} solution to the above quadratic equation.

Denote $t = T_{ij}$, $f = \mathcal{F}_{ij}$, $T_1 = \min\{T_{i-1,j}, T_{i+1,j}\}$, and $T_2 = \min\{T_{i,j-1}, T_{i,j+1}\}$. Equation (7.2) reads

$$(\max\{t - T_1\}, 0\})^2 + (\max\{t - T_2\}, 0\})^2 = f^2,$$

where T_1, T_2, and f are given and t is the unknown. We have the following cases for t.

- $t > \max\{T_1, T_2\}$, for which t is the bigger solution of the quadratic equation

$$2t^2 - 2t(T_1 + T_2) - f^2 + T_1^2 + T_2^2 = 0,$$

  that is,

$$t = \frac{T_1 + T_2 + \sqrt{2f^2 - (T_1 - T_2)^2}}{2}.$$

- $T_2 > t > T_1$, in which case we have the solution $t = T_1 + f$. This case happens only if $T_2 - T_1 > f$.

- $T_1 > t > T_2$, in which we have the dual solution $t = T_2 + f$. This happens only if $T_1 - T_2 > f$.

The final numerical scheme involves an initialization, in which for all vertices $u = \{ij\}$ we set $T_u = \infty$, and for all vertices $v = \{kl\}$ where initial/boundary conditions are given, we fix $T_v =$given value. The update procedure for a grid point (vertex) $u = \{ij\}$ is given by the following simple procedure that computes T_u, and either inserts it into a priority queue Q, or updates its location in Q.

Update(u, Q)

$\{i, j\} = u$
$T_1 = \min\{T_{i-1,j}, T_{i+1,j}\}$
$T_2 = \min\{T_{i,j-1}, T_{i,j+1}\}$
if $|T_1 - T_2| < \mathcal{F}_{ij}$ then
$$t = \frac{T_1 + T_2 + \sqrt{2\mathcal{F}_{ij}^2 - (T_1 - T_2)^2}}{2}$$
else $t = \min\{T_1, T_2\} + \mathcal{F}_{ij}$
$T_{ij} = \min\{T_{ij}, t\}$
if $u \in Q$ then $Reorder(u, Q)$
  else $Q \leftarrow Q \cup u$

Note that the value of an updated point is always bigger than the values of the points that contribute to its update. This observation, coupled with the initialization to ∞, makes this monotone scheme consistent, and suggests the order of the update events. The flow of information is always "upwards" from low to high values of T. By ordering the updates, the viscosity solution is obtained with computational complexity bounded by $O(M \log M)$ on a sequential machine, where M is the number of grid points.

Similar to the Dijkstra algorithm, the fast marching algorithm goes as follows. Define the group S that contains all vertices for which the T vlaue is decided, and G as the group of all vertices. That is, $\forall \{i, j\} \in S$, T_{ij} either satisfies Eq. (7.2) or is given as a fixed boundary condition. The algorithm updates all the adjacent vertices of S. First the T values of all the adjacent vertices are entered to a priority queue called Q. Each main step of the algorithm, a vertex $u = \{i, j\}$, whose T_{ij} is the smallest, is extracted from Q and moved to S, and all its neighboring vertices that are not in S are updated. An updated vertex that changed its value can either enter the queue Q if it is the first time it is updated, or change its location in Q if it is already there. The queue, Q, can be implemented as a binary heap (see, e.g., [186]). The following pseudocode explains the algorithm:

Fast Marching($\mathcal{F}, G, S, Q$)

$\forall v \in G \setminus S$ do $T_v \leftarrow \infty$
while $Q \neq \emptyset$
    do $u \leftarrow ExtractMin(Q)$
      $S \leftarrow S \cup u$
      for each vertex $v \in Adj[u]$
        do $Update(Q, v)$

An alternative explanation is following Sethian's notations in [190, 191, 192], in which each grid point is tagged as either *Alive* (in S), *Close* (in Q), or *Far*. All points with given initial values are set to be *Alive*, which is the initial set of accepted values. All points one grid point away from the *Alive* points are initialized to be *Close*, which is an ordered list of candidate points. The rest of the points are set to be *Far*. The fast marching method sequentially moves points from the set *Far* to the set of accepted values *Alive*, through the front set *Close*. See Figure 7.5.

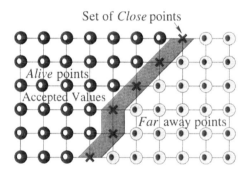

Figure 7.5: The ordered set *Close* is the front between the *Far* points, with ∞ values, and *Alive* points, with accepted values.

The solution T is systematically constructed from smaller to larger values of T. A point gets updated only by points with smaller values. This "monotone property" allows us to keep a front of candidate points, that tracks the flow of information, ordered in a heap structure in which the root is always the smallest value. An update of an element in the heap with M elements is done in $O(\log M)$. Thereby, an upper bound of the total computational complexity is $O(M \log M)$.

The method includes the same ingredients as Dijkstra's method [57, 186] that finds paths of minimum cost in graphs. Yet, Dijkstra's algorithm would fail to consistently solve geometric problems. Actually, any graph search-based algorithm would fail to converge as the graph resolution is systematically refined. The graph induces an artificial metric that would be inconsistent with the continuous case.

7.2.1 First Application: Path Planning

Robot motion planning addresses the question of how should a robot change its position between two given configurations. Based on [111] and motivated by [97, 106], we deal with the problem of path planning for a robot with a small number of degrees of freedom (DOF), navigating in weighted domains. Classically, graph search-based methods are applied to these type of problems, in which case the graph imposes an unnatural metric on the physical continuous problem. Resolution refinement of the graph does not improve the errors that characterize these methods. As an attractive consistent replacement with a similar computational complexity, the fast marching method can be used for this task.

Let us start with a simple example of a point robot navigating on a flat plane. In this case, the Euclidean geometry provides us with a straightforward solution to the optimal path problem: The straight line connecting the two configurations is the optimal path. Let us first try to solve this simple problem, using a simple graph search-based method, say Dijkstra's method [57]. Assume we use a rectangular grid on the unit square and measure the distance along the horizontal and vertical edges connecting the grid points; see Figure 7.6. The length of the optimal path between the lower left point $(0,0)$ and the upper right point $(1,1)$ will always be 2 according to graph search measurements on the given network. No matter how much we refine the mesh resolution, going along the edges of the square will be as optimal as any other left-to-right top-to-bottom sequence that connects the two configurations. In fact, the network imposes the L_1 metric on the problem, in which the "unit circle" has a diamond shape.

Latombe's book [127] is a nice collection of motion planning algorithms. Indeed, for a small number of dimensions, say up to six degrees of freedom, most of the classical planners used graph-based methods. For the many degrees of freedom planning problems, due to the "curse of dimensionality" one usually gives up optimality and searches for computational efficiency. In some cases, heuristic functions are used to solve the navigation task. Other methods sample the configuration space and try to link between the samples and construct a "road map" in the high-dimensional configuration space. In the following application we limit our discussion to the class of problems involving a small number of DOF. These are the interesting cases for which the geometric-numerical fast marching method is a better alternative.

Figure 7.7 shows six examples of path planning for rectangular shapes that translate and rotate in a given cluttered environment. The 2D work space includes stationary obstacles, and the question is how to navigate

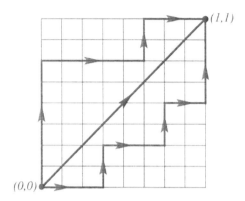

Figure 7.6: The graph search optimal results will never converge to the diagonal true solution.

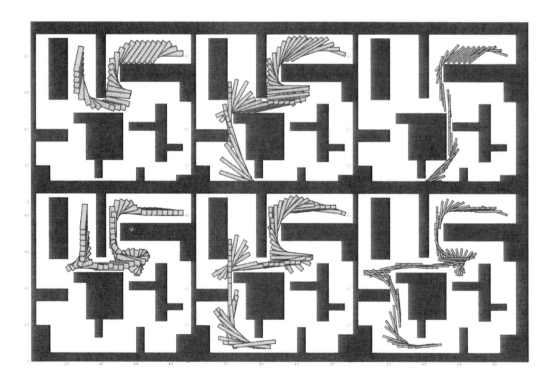

Figure 7.7: Path planning for a rectangular shape with translation and rotation. The lower frame of each couple shows the influence of a penalty of the distance from the obstacles on the path.

the rectangle from its given initial position to a given final position. The solution is a simple application of the fast marching method in the $\{x, y, \psi\}$ configuration space. First, the 3D configuration space is constructed. For each discrete angle $k\Delta\psi$, we apply a morphological dilation of the obstacles with the rectangle rotated by $k\Delta\psi$ as a structuring element. This way we reduce the problem to a path planning problem for a point robot, in a configuration space whose dimensionality is determined by the number of degrees of freedom (three in this example). Next, we use the fast marching method to compute the distance from the initial configuration to each point in the free configuration space. This is done by solving the eikonal equation

$$\sqrt{(\partial_x T)^2 + (\partial_y T)^2 + (\partial_\psi T)^2} = \mathcal{F}_{ijk},$$

given $T(x_0, y_0, \psi_0) = 0$, where, for example, $\mathcal{F}_{ijk}$ is an increasing function of the distance from the obstacles in the $\{x, y, \psi\}$ space. Finally, the optimal path is extracted by back propagation along the gradient direction of the computed distance map. This is the minimal geodesic with respect to the arclength $ds^2 = \mathcal{F}^2(x, y, \psi)(dx^2 + dy^2 + d\psi^2)$, for this navigation problem.

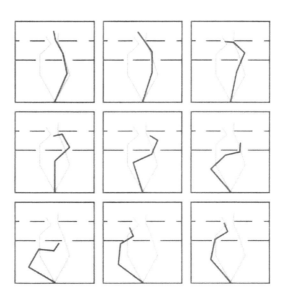

Figure 7.8: Path planning for an arm with four degrees of freedom. Snapshots along the optimal path from the initial to the final configuration are shown left to right, top to bottom.

Figure 7.8 presents a solution of a path planning problem for a robot arm with four degrees of freedom. Now the eikonal equation, in the $\{\psi_1, \psi_2, \psi_3, \psi_4\}$ configuration space, reads

$$\sqrt{(\partial_{\psi_1} T)^2 + (\partial_{\psi_2} T)^2 + (\partial_{\psi_3} T)^2 + (\partial_{\psi_4} T)^2} = \mathcal{F}_{ijkl}.$$

Again, we apply the fast marching method followed by a back propagation via the distance gradient direction to extract the shortest path. The path is the shortest path with respect to the arclength measure $ds^2 = \mathcal{F}^2(\psi_1, \psi_2, \psi_3, \psi_4)(d\psi_1{}^2 + d\psi_2{}^2 + d\psi_3{}^2 + d\psi_4{}^2)$, for this navigation problem.

7.3   Fast Marching on Triangulated Manifolds

In [112], Sethian's fast marching method was extended to triangulated domains. Upwind operators for level set schemes on triangulated domains were developed in [10]. The extension of the fast marching method is based on the observation that a first-order numerical method constructs a piecewise planar surface as a solution for $|\nabla T_{ij}| = \mathcal{F}_{ij}$. That is, the update procedure may be viewed as fitting a tilted plane with gradient defined by the value of $\mathcal{F}_{ij}$ at one grid point, and two values anchored at the relevant neighboring grid points. Since three points define a plane, the numerical update procedure may be extended to a vertex of a triangle with given values at its other vertices. Let us start with acute triangles and then provide a special treatment for obtuse angles.

Considering the solution at a specific vertex, one should consider the possible contribution for an update value from all triangles that share that vertex. We choose the one that produces the smallest new value for T.

The update procedure for a single vertex in a given acute triangle goes as follows. Consider the nonobtuse triangle ABC in which the point to update is C. For consistency and monotonicity, the update should be from within the triangle. Otherwise, decreasing the value of an updating vertex may increase the value of the updated vertex, with a contradiction to the monotonicity requirement. We should thus verify that the altitude h is inside the triangle ABC. See Figure 7.9.

We search for $t = EC$ that satisfies the gradient approximation

$$\frac{t - u}{h} = \mathcal{F}, \tag{7.3}$$

where we assume w.l.o.g. that $T(A) = 0$. The above relation sets the slope of the plane EHA to be $\mathcal{F}$. It is just one part of the whole distance map.

Denote $a = BC$ and $b = AC$; by similarity we have $t/b = DF/AD = u/AD$, thus

$$CD = b - AD = b - bu/t = b(t - u)/t. \tag{7.4}$$

By the law of cosines,

$$BD^2 = a^2 + CD^2 - 2a\,CD\cos\theta,$$

and by the law of sines,

$$\sin\phi = \frac{CD}{BD}\sin\theta.$$

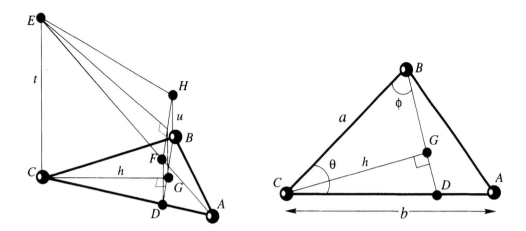

Figure 7.9: Given the triangle ABC, such that $u = T(B) - T(A)$, find $T(C) = T(A) + t$ such that $(t - u)/h = \mathcal{F}$. Left: A perspective view of the triangle stencil supporting the $T()$ values, which form a tilted plane with a gradient magnitude equal to $\mathcal{F}$. Right: The trigonometry on the plane defined by the triangle stencil.

Using the right angle triangle CBG, we have

$$h = a \sin \phi = a \frac{CD}{BD} \sin \theta = \frac{a\, CD\, \sin \theta}{\sqrt{a^2 + CD^2 - 2a\, CD \cos \theta}}. \qquad (7.5)$$

Next, we plug the relation (7.4) into (7.5), and the result into the gradient approximation equation (7.3). We end up with the quadratic equation for t,

$$c^2 t^2 + 2bu(a \cos \theta - b)t + b^2(u^2 - \mathcal{F}^2 a^2 \sin^2 \theta) = 0, \qquad (7.6)$$

where $c^2 = AB^2 = a^2 + b^2 - 2ab \cos \theta$.

The solution t must satisfy $u < t$, and should be updated from within the triangle, namely

$$a \cos \theta < \frac{b(t - u)}{t} < \frac{a}{\cos \theta}. \qquad (7.7)$$

Thus, the update procedure is given by the following procedure:

if $u < t$ and $a \cos \theta < \frac{b(t-u)}{t} < \frac{a}{\cos \theta}$
  then $T(C) = \min\{T(C), t + T(A)\}$
   else $T(C) = \min\{T(C), b\mathcal{F} + T(A), a\mathcal{F} + T(B)\}$

This equation is a finite difference approximation to the eikonal equation on the triangle. It is monotone by construction, consistent, converges to the viscosity solution, and is thereby a valid extension of Sethian's fast marching method to acute triangulated domains.

Let us consider the following example of an obtuse angle. Let the coordinates of the three vertices of a triangle in a plane be $A = (0, 0)$, $B = (1, 0)$, and $C = (-1, 0.5)$, and the point to be updated is A. Now, assume a wavefront coming from $(1, 1)$ (see Figure 7.10), where the front is the line $y(t) = (2 - t) - x$, approaching the origin.

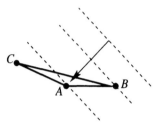

Figure 7.10: Example of difficulty in obtuse angles. An equal-distance contour line approaches A without touching its neighbor C.

This front, or equal-distance line, would first meet the point B ($t = 1, y = 1 - x$) and then the point A ($t = 2, y = -x$), and only then the point C. We see that the point A is "supported" by a single point that cannot recover the actual direction of the coming front. Actually, the only supported section of incoming fronts is a limited section, as shown in Figure 7.12. We can extend this section and link the vertex to another one within the extended section and thereby split the angle and exclude such problems.

As we just saw, in case of an obtuse angle the support may include only a limited section of incoming wavefronts. One approach to overcome such problems is to construct a local numerical support at obtuse angles by splitting these angles. Connecting the vertex to any point in the limited supported section splits the obtuse angle into two acute ones. The idea is to extend this section by recursively unfolding the adjacent triangles, until a new vertex Q is included in the extended section. Then, the vertices are connected by a virtual directional edge from Q to A (i.e., A may be updated by Q). The length of the edge AQ is equal to the distance between A and Q on the unfolded triangles plane. See Figure 7.11.

Let us give a complexity analysis for the node capturing. Let $h_{\max}, h_{\min}$ be the maximal and minimal altitudes, respectively, that is, triangles altitudes with maximal and minimal length. Let $\theta_{\max}$ be the maximal obtuse angle, denote $\alpha = \pi - \theta_{\max}$ the angle of the extended section, and let $\theta_{\min}$ be the minimal (acute) angle for all triangles. Let $e_{\max}$ be the length of the longest edge, and let l be the length of the virtual directional edge (AQ

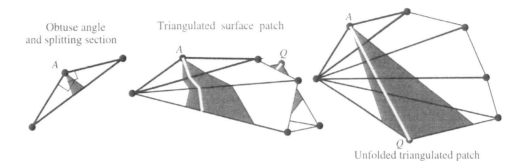

Figure 7.11: Left: The initialization of the construction for the splitting section. Middle: A triangulated surface patch. Right: The unfolded patch and the splitting section expansion up to the first vertex Q, and the virtual edge connecting the two vertices AQ.

in the above example). Furthermore, assume that α and $\theta_{\min}$ are small so that $\sin \alpha \approx \tan \alpha \approx \alpha$.

Then, the angular width of the narrower section is $\alpha = \pi - \theta_{\max}$, so that we have $\sin \alpha/2 \leq e_{\max}/2l$. This relation, for small α angles, yields $l \leq e_{\max}/\alpha$. Denote by $l_{\max} = e_{\max}/\alpha$.

The maximal area of the extended sections is bounded from above by

$$a_{\max} = \frac{e_{\max}^2}{2\alpha},$$

while the minimal area of an unfolded triangle is bounded from below by

$$a_{\min} = \frac{(h_{\min}\alpha)^2 \theta_{\min}}{2};$$

see Figure 7.12. Therefore, the number of triangles that need to be unfolded before a vertex is found in the extended section is bounded by the ratio of these areas:

$$m = \frac{a_{\max}}{a_{\min}} = \frac{e_{\max}^2}{\theta_{\min} h_{\min}^2 \alpha^3}. \tag{7.8}$$

The accuracy of the first-order scheme for acute triangles is of $O(h_{\max}) \approx O(e_{\max})$, while the accuracy for the obtuse case with the above construction becomes $O(l_{\max}) = O(e_{\max}/(\pi - \theta_{\max}))$. As expected, in the worst-case scenario, the scheme accuracy depends on the largest edge and the widest angle.

The construction of the virtual directional edges includes unfolding triangles for each obtuse angle until a vertex is detected in the extended splitting section. Since the number of unfolded triangles is bounded by a

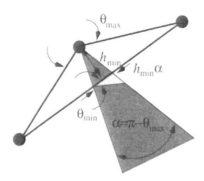

Figure 7.12: The smallest possible area of one triangle covered by the longest possible extended section bounds the number of unfolded triangles. The triangle with $a_{\min}$ is the bright shaded one.

constant, the construction of the virtual directional edges takes $O(M)$, and the total computational complexity is still $O(M \log M)$.

7.4 Applications of Fast Marching on Surfaces

An efficient method for distance computation on triangulated surfaces enables us to solve interesting problems. In the first example, taken from [112], we search for the minimal geodesics connecting two points on a curved surface. A minimal geodesic is the shortest path connecting the two points, where the path is restricted to lay on the surface. That is, no shortcuts are allowed. Next, following [114], we apply the fast marching method on triangulated domains to compute offset contours and Voronoi diagrams on surfaces.

7.4.1 Minimal Geodesics

Figure 7.13 presents two surfaces with one point from which the geodesic distance is first computed. Next, a local gradient descent process is applied to integrate minimal geodesics from different locations. The minimal geodesics are extracted from a given initial point by back propagating along the distance map. There are many numerical schemes to integrate the path. One simple method uses a linear interpolation of the path at each triangle. That is, given a point along one of the edges of the triangle, compute a line segment along the gradient direction of the distance map defined at the vertices of the triangle. See Figure 7.14.

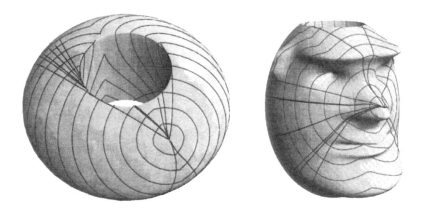

Figure 7.13: A perspective view of the shortest paths and equal-distance contours from a point on a bead and a point at the tip of the nose of a synthetic head.

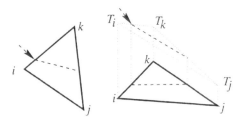

Figure 7.14: Extracting a segment of a minimal geodesic by linear interpolation. Left: A triangle and the linear interpolation of a minimal geodesic, both embedded in the figure plane. Right: A perspective view of the triangle and the geodesic distance values at its vertices at the third dimension.

Figure 7.15 presents a cubic polynomial interpolation of the distance map. The idea here is to unfold the three neighboring triangles around the one that is under inspection. Each triangle contributes a vertex with its distance value that supports the surface interpolation. In this case, one can use a higher-order procedure to extract the minimal geodesic. One example is Heun's method. Given the planar O.D.E.

$$\partial_t C = -\nabla T \qquad \text{given} \qquad C(0),$$

Heun's method is a two-step, second-order integration method that reads as follows:

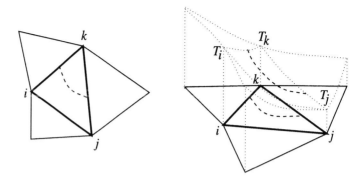

Figure 7.15: Extracting a segment of a minimal geodesic by cubic interpolation of the distance map and second-order integration of the path. Left: The current triangle and its three neighboring triangles, unfolded to the figure plane. Right: A perspective view of the unfolded neighboring triangles, the interpolation polynomial geodesic distance as the third dimension, and the minimal geodesic segment.

- Initialization: Let $p_0 = C(0)$ be the given initial point, and initialize $p_2 = C(0)$.

- While p_2 is within the support of the current triangle, repeat:

- Step 1:

$$\vec{V_0} = -\nabla T(p_0),$$
$$p_1 = p_0 + dt\vec{V_0}.$$

- Step 2:

$$\vec{V_1} = -\nabla T(p_1),$$
$$p_2 = p_0 + dt(\vec{V_1} + \vec{V_0})/2.$$

- Draw the segment (p_0, p_2), let $p_0 = p_2$.

- END of while loop.

See Figure 7.16.

$\nabla T(p)$ is evaluated by taking the derivatives of the second-order interpolation polynomial at the coordinate position p. The sequence of segments (p_0, p_2) is the desired geodesic path within the triangle. Moving from one triangle to another involves the construction of a new interpolation polynomial for T. The whole process is repeated until $T(p_2) \approx 0$.

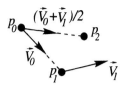

Figure 7.16: Heun's second-order integration method is applied to back track to the shortest path.

7.4.2   Fast Voronoi Diagrams and Offsets on Surfaces

In [114] the fast marching method on triangulated domains is used to efficiently compute Voronoi diagrams and offset curves on triangulated manifolds. The computational complexity of the algorithm is again $O(M \log M)$, where M is the number of vertices in the triangulated surface. The algorithm also applies to weighted domains in which a different cost is assigned to each surface point.

Voronoi diagrams play important roles in many fields such as robotic navigation, image processing, computer graphics, pattern recognition, computer vision, and more. Its flat Euclidean version, for which there is an efficient implementation known as Fortune's algorithm [72], is a building block in many applications.

The Voronoi diagram sets boundaries between a given set of source points and splits the domain into regions such that each region corresponds to the closest neighborhood of a source point from the given set. Let our domain be D, let the set of given n points be $\{p_j \in D, j \in 0, .., n-1\}$, and let the distance between two points $p, q \in D$ be $d(p, q)$; then the Voronoi region G_i corresponds to the set of points $p \in D$ such that $d(p, p_i) < d(p, p_j), \forall j \neq i$. See Figure 7.17, left.

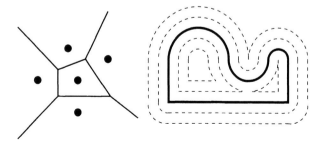

Figure 7.17: Left: Voronoi diagram of five points on the plane. Right: Offset curves of a simple planar curve.

Offset computation is often used in approximation and singularity theories and comes into practice in computer aided design (CAD) and the numerical control industry (e.g., in NC machines). Given a curve and its embedding space, an offset curve is defined by a set of points with a given fixed distance from the original curve. See Figure 7.17, right.

There are some numerical and topological difficulties even in the computation of offsets for curves in the 2D Euclidean plane, for instance the formation of singularities in the curvature, self-intersection of the offsetting curve, and the fact that an offset of a polynomial parameterized curve is not necessarily polynomial. Some of the numerical difficulties are addressed in [99], where the Osher–Sethian level set method [161, 191] is used to overcome the topological changes.

Efficient construction of distance maps, minimal geodesics, Voronoi diagrams, and offset curves for curved and weighted domains is a challenging problem; see, for example, [149, 141, 97, 124, 66, 105]. It has many potential applications like path planning for a milling tool covering a given surfaces in CAD, texture mapping in computer graphics, and intrinsic coordinate generation. Let us apply the fast marching method to solve these problems. The proposed algorithm's computational complexity is again $O(M \log M)$, its implementation is simple, and it applies to weighted domains in which a different cost is assigned to each surface point.

We have an algorithm to compute distances on triangulated manifolds, and hence construct offset curves. First, we solve the eikonal equation with speed $\mathcal{F} = 1$ on the triangulated surface to compute the distance from a source region that defines the initial curve. We then find the equal geodesic distance curves on the surface by interpolating the intersection with a constant threshold using a "marching triangle" procedure, again an $O(M)$ operation. The marching triangle procedure constructs a piecewise linear approximation for a level set curve of a function defined on the surface by considering its values at the vertices. Say we search for the zero level set of T. Given a triangle with the function values at its three vertices T_1, T_2, T_3, if $\max\{T_1, T_2, T_3\} < 0$ or $\min\{T_1, T_2, T_3\} > 0$, then the curve does not intersect with the triangle. Otherwise, we linearly interpolate the two intersection points of the triangle edges with the curve and produce a line segment by connecting these points. The set of all segments forms an approximation to the zero set curve of the function T on the surface.

Figure 7.18 shows two examples of offset computation and Voronoi diagram for five points on two surfaces, while in Figure 7.19, we introduce a different weight at each point of the surface. The gray level of the surface in this case is proportional to its relative weight.

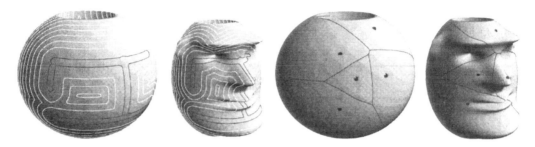

Figure 7.18: Offsets and Voronoi diagrams of five points on a bead and a synthetic head.

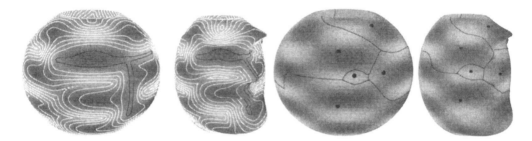

Figure 7.19: Offsets and Voronoi diagrams of five points on a bead and a synthetic head with weights.

7.5   Exercises

1. Research problem: Define and test mathematical morphology on smooth manifolds. Extend and apply the fast marching on triangulated domains to compute opening and closing.

2. Research problem: Extend the fast marching on triangulated domains to 3D triangulations. Work out the cubic root algorithm complexity such that an $O(\Delta x)$ accurate solution is computed in constant time for the update stage.

3. Implement a path planner for a robot navigating between obstacles. The robot is a rectangle that rotates and translates. An optimal path between the initial and final configurations should be the result of a minimal path $C(s) = \{x(s), y(s), \psi(s)\}$ weighted by an inverse distance from the obstacles in the x, y, ψ configuration space. Denote the $\{x, y, \psi\}$ configuration space as $\mathcal{CS}$, the free space as $\mathcal{FS}$, and the

obstacle space as $\mathcal{OS}$. Then use the inverse distance function in free object space $\mathcal{FS}$,

$$F(p \in \mathcal{FS}) = \sup_{q \in \mathcal{OS}} \frac{1}{1 + \mathrm{d}_e^2(p, q)},$$

and $F(p \in \mathcal{OS}) = \infty$, and find the path that minimizes the functional

$$\min \int F(C(s)) ds,$$

where $ds^2 = dx^2 + dy^2 + d\psi^2$ is the Euclidean arclength in the x, y, ψ configuration space. Construct a consistent solution, and explain your choice of algorithm and its complexity for each step in the solution. Use a morphological approach for the construction of the obstacle space. Use an $O(M)$ approach for the distance computation, then an $O(M \log M)$ approach based on the fast marching method for the weighted distance computation, followed by a steepest descent subgrid integration method for back tracking the optimal path. Bonus: Apply your algorithm to any example with seven DOF.

4. Apply the fast marching method to compute distances, minimal geodesics, offsets, and Voronoi diagrams, on a Klein Bottle in $\mathbb{R}^4$, and a Möbius Strip in $\mathbb{R}^3$. See [58] for parametric forms of these two surfaces.

Shape from Shading

One of the earliest problems in computer vision is the "shape from shading problem" [87, 89, 88]. The question is how to reconstruct the shape of a smooth "Lambertian" object from a given single gray-level image. Excluding interreflections and specularities, the shading image of a Lambertian object is a function of the projection of the surface normal onto the light source direction. The question then is how to extract the surface normal direction at each point, and integrate the normal vector field into a valid surface. The main problem involves the search for necessary and meaningful conditions for existence, uniqueness, and the design of a consistent, stable, efficient and accurate numerical scheme that would reconstruct the shape of the object.

Even nowadays, computer vision books and courses introduce solutions to the shape from shading problem based on variational principles that require additional smoothness or additional regularization terms. This requirement introduces unnecessary second-order derivatives into the minimization process; see [89]. Two direct models for solving the shape from shading did not incorporate extra smoothness terms: The first is the characteristic strips expansion method used by Horn when he first introduced the problem [87]; the second is Bruckstein's equal-height contours tracking model [25]. Unfortunately, direct numerical implementations of these algorithms suffered from numerical instabilities.

Modern numerical algorithms based on recent results in curve evolution theory, control theory, and the viscosity solution framework [54] were recently applied to the shape from shading problem [170, 61, 13, 98, 101]. In these numerical methods the smoothness assumption is part of the scheme without the need for extra smoothness terms as a penalty.

Part of the reconstruction problem is the extraction of the surface topology, or more accurately the surface topography. One problem is that a saddle, a local maximum, and a local minimum point may have the same local shading image. It is shown in [100] that the reconstruction problem can be uniquely solved for smooth surfaces with complicated topologies as long as the surface normals are known to point outward along the boundaries of a given domain like the image boundaries; see [24] for a related effort.

8.1   Problem Formulation

Consider a smooth three-dimensional object defined by $S(x, y) = \{x, y, z(x, y)\}$, where $z(x, y) : \mathbb{R}^2 \to \mathbb{R}$, and a single parallel light source at infinity defined by its direction $\vec{l} = \{l_1, l_2, l_3\}$. The surface normal is given by

$$\vec{\mathcal{N}} = \frac{S_x \times S_y}{|S_x \times S_y|} = \frac{\{-z_x, -z_y, 1\}}{\sqrt{1 + z_x^2 + z_y^2}}.$$

The gray-level shading image of a *Lambertian* object for an observer direction $\vec{v} = \{0, 0, 1\}$ is defined by

$$I(x, y) = \lambda \langle \vec{\mathcal{N}}, \vec{l} \rangle = \lambda \frac{-z_x l_1 - z_y l_2 + l_3}{\sqrt{1 + z_x^2 + z_y^2}},$$

where λ is a constant proportionality factor that can be set to 1 by rescaling the image. See Figure 8.1.

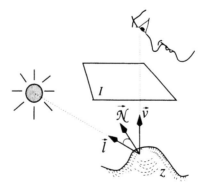

Figure 8.1: The shading image I of a Lambertian object is given by $I = \langle \vec{\mathcal{N}}, \vec{l} \rangle$.

Shape from shading refers to the reconstruction problem of the three-dimensional shape $z(x, y)$ from the given shading image $I(x, y)$. Let us start with the simple case where $\vec{l} = \vec{v} = \{0, 0, 1\}$, in which the shading image is given by

$$I(x, y) = \frac{1}{\sqrt{1 + z_x^2 + z_y^2}}.$$

This equation can be compactly written as $I = (1 + |\nabla z|^2)^{-1/2}$, or as an *eikonal* equation

$$|\nabla z| = F(x, y),$$

where in our case

$$F(x, y) = \frac{\sqrt{1 - I^2(x, y)}}{I(x, y)}. \tag{8.1}$$

The points at which $I = 1$ (i.e., $|\nabla z| = 0$) define the extremum and saddle points of the function z. Note, that both $z_1(x, y) = \frac{1}{2}(x^2 + y^2)$ and $z_2(x, y) = \frac{1}{2}(x^2 - y^2)$ share the same shading images, so there is no local way to distinguish among a minimum, a maximum, and a saddle. See Figure 8.2.

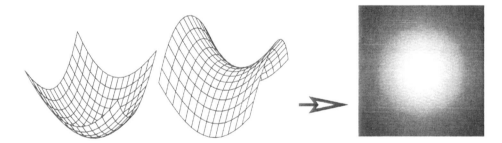

Figure 8.2: The surfaces, $z_1(x, y) = \frac{1}{2}(x^2 + y^2)$ and $z_2(x, y) = \frac{1}{2}(x^2 - y^2)$, with light source at the viewer direction $\vec{l} = \vec{v} = \{0, 0, 1\}$, generate the same shading image $I(x, y) = 1/\sqrt{1 + x^2 + y^2}$.

8.2 Horn Characteristic Strip Expansion Method

Assume we know the local structure of the surface along a given curve in the image plane. Say, we know z, z_x, and z_y at the point (x_0, y_0). Then Horn's idea is to integrate the surface as a solution along *characteristic strips*. Let us define the intensity image again as a function of a *reflectance* map $R(\vec{N}, \vec{l}, \vec{v})$. For the simple Lambertian case with light at the viewer direction

$$I(x,y) = R(\vec{N}) = R(z_x(x,y), z_y(x,y)),$$

note that by the chain rule we have that

$$
\begin{aligned}
dz &= z_x dx + z_y dy, \\
dz_x &= z_{xx} dx + z_{xy} dy, \\
dz_y &= z_{xy} dx + z_{yy} dy, \\
I_x &= R_{z_x} z_{xx} + R_{z_y} z_{xy}, \\
I_y &= R_{z_x} z_{xy} + R_{z_y} z_{yy}.
\end{aligned}
$$

We now use our freedom for choosing (dx, dy), such that

$$
\begin{aligned}
dx &= R_{z_x} dw, \\
dy &= R_{z_y} dw,
\end{aligned}
$$

which yields

$$
\begin{aligned}
dz_x &= z_{xx} dx + z_{xy} dy \\
&= z_{xx} R_{z_x} dw + z_{xy} R_{z_y} dw = I_x dw,
\end{aligned}
$$

and similarly

$$dz_y = I_y dw.$$

We can now collect a system of five ordinary differential equations that describe the path of one characteristic on the surface z,

$$
\begin{aligned}
dx &= R_{z_x} dw, \\
dy &= R_{z_y} dw, \\
dz &= (z_x R_{z_x} + z_y R_{z_y}) dw, \\
dz_x &= I_x dw, \\
dz_y &= I_y dw.
\end{aligned}
$$

Denote $p \equiv z_x$, and $q \equiv z_y$, which allows us to write the above system of ODEs as

$$
\frac{d}{dw}
\begin{pmatrix} x \\ y \\ z \\ p \\ q \end{pmatrix}
=
\begin{pmatrix} R_p \\ R_q \\ pR_p + qR_q \\ I_x \\ I_y \end{pmatrix}.
$$

A simple numerical approximation yields

$$
\begin{aligned}
x^{n+1} &= x^n + R_p^n \Delta w, \\
y^{n+1} &= y^n + R_q^n \Delta w,
\end{aligned}
$$

$$
\begin{aligned}
z^{n+1} &= z^n + (p^n R_p^n + q^n R_q^n)\Delta w, \\
p^{n+1} &= p^n + I_x \Delta w, \\
q^{n+1} &= q^n + I_y \Delta w,
\end{aligned}
$$

where $\{x^n, y^n, z^n\}$ is the nth point along the surface path. Horn suggested to start from small circles around the local maximum points and expand the information "downward." See Figure 8.3.

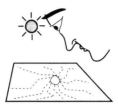

Figure 8.3: The characteristic strip expansion method starts from a given curve with known height and gradient and expands the solution by tracking the characteristics of a system of five ODEs.

The main problem with this approach is that each characteristic has its own "life," and thus two may accidentally meet due to numerical errors. One can design stability procedures that control the creation of new characteristics and termination of others, yet this leads to nonnatural and complicated schemes.

8.3 Bruckstein's Equal-Height Contours Expansion Method

Bruckstein suggested a different direct approach based on incremental construction of the surface level sets or equal-height contours [25]. Let's assume we are given one contour $C(0)$ along which $z(x, y)$ is a constant, for example $z = 0$. Denote $t \equiv z$, that is, we refer to the depth as the time it takes the level set to get to a specific value. Then $z_t = 1$ and we can use the chain rule

$$
\begin{aligned}
z_t &= z_x x_t + z_y y_t \\
&= \langle \nabla z, C_t \rangle.
\end{aligned}
$$

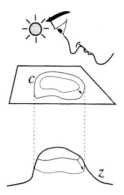

Figure 8.4: The equal-height contour evolution equation is given by $C_t = (I/\sqrt{1 - I^2})\vec{N}$,

Dividing both sides of the last equation by $|\nabla z|$, we obtain

$$\frac{1}{|\nabla z|} = \left\langle \frac{\nabla z}{|\nabla z|}, C_t \right\rangle$$
$$= \langle \vec{N}, C_t \rangle.$$

We already know that normal to the level set C of z coincides with $\nabla z/|\nabla z|$; we thereby obtain a propagation rule for the equal-height contours given by

$$C_t = \frac{1}{|\nabla z|}\vec{N}.$$

Note that $\vec{N}$ is the planar normal of the planar curve C. For the simple shape from shading case studied in Eq. (8.1), we have

$$C_t = \frac{I(x, y)}{\sqrt{1 - I^2(x, y)}}\vec{N} \qquad \text{given} \qquad C(t = 0). \qquad (8.2)$$

See Figure 8.4. Like the characteristic strip expansion method, direct numerical implementation of the equal-height contour evolution suffers from numerical instabilities. Problems occur as curvature singularities appear in the propagating contour, or as a single connected contour changes its topology into two separate curves.

8.4   Tracking Level Sets by Level Sets

In order to overcome the numerical difficulties and handle the topological changes of the propagating equal-height contour, the level set approach was applied to the problem [92, 98, 101, 116]. We rewrite Eq. (8.2) in its level set form, which reads

$$\phi_t = \frac{I(x,y)}{\sqrt{1 - I^2(x,y)}}|\nabla\phi|.$$

See Figure 8.5.

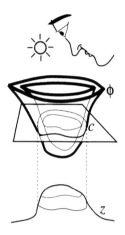

Figure 8.5: The equal-height contour evolution equation in level set formulation reads $\phi_t = (I/\sqrt{1 - I^2})|\nabla\phi|$.

The numerical implementation is based on the monotone scheme described in Chapter 5, given by

$$\phi_t = \frac{I}{\sqrt{1 - I^2}}\sqrt{(\max\{-D^x_+\phi, D^x_-\phi, 0\})^2 + (\max\{-D^y_+\phi, D^y_-\phi, 0\})^2}.$$

Observe that if we could find a ϕ function such that

$$|\nabla\phi|\frac{I}{\sqrt{1 - I^2}} = 1,$$

then $\phi(x,y) = z(x,y) + \text{constant}$, which basically solves the problem. That is, we are back with the eikonal equation

$$|\nabla z| = \frac{\sqrt{1 + I^2}}{I}.$$

Rouy and Tourin [170] suggest to directly solve the monotone discretization of this equation,

$$\frac{\sqrt{1 - I^2}}{I} = \sqrt{(\max\{-D^x_+z, D^x_-z, 0\})^2 + (\max\{-D^y_+z, D^y_-z, 0\})^2}.$$

By its monotonicity, this numerical scheme selects the viscosity solution. Rouy and Tourin suggest to scan the image domain and solve the resulting quadratic equation for each grid point. In Chapter 7 we saw that by ordering the updates, z can be reconstructed very efficiently using the fast marching method.

8.5   Extracting the Surface Topography

The above schemes are useful for computing the solution from a single given source point upward or downward. That is, from a given maximum point the reconstructed solution is always decreasing as the equal-height contour propagates downward. However, tracking the level sets of the surface enables us to reconstruct more complicated functions. In fact, by limiting the class of surfaces to *Morse* functions, we are able to extend local solutions and reconstruct complicated surfaces as shown in [100].

Morse functions are smooth functions in which pathological geometries like monkey saddles are excluded (see Figure 8.6), that is, all points at which $|\nabla z| = 0$ can be mapped by a local affine transformation into $d \pm x^2 \pm y^2$.

Figure 8.6: A pathological case known as the monkey saddle.

An equal-height contour that starts its downward propagation at a local maximum is always decreasing up to a saddle point where it is connected to another contour. A simplification of Euler–Poincaré characteristics and Poincaré's theorem known as the Mountaneer's theorem states that the number of extremum points within an equal-height contour is equal to the number of saddle points plus one.

$$\#maximum + \#minimum - \#saddle = 1.$$

We can distinguish between two types of saddle points based on the way the equal-height contour reaches the point. The "first type" is characterized by the fact that the same equal-height contour reaches the saddle from two opposite directions and splits into two simple equal-height contours that continue decreasing. See Figure 8.7, top. The "second type" reaches the saddle point from one direction. It should "wait" for another equal-height contour to reach the saddle from its opposite direction, merge with the second contour, and continue the downward propagation as a single equal-height contour. See Figure 8.7, bottom.

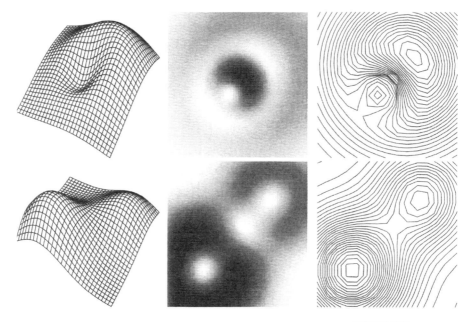

Figure 8.7: The surface z, its shading image $I = 1/\sqrt{1 + |\nabla z|^2}$, and the equal-height contours, for the two types of saddles.

Assume the surface normals are pointing outward along the boundary of the image domain or the object boundary. The following simple algorithm reconstructs the topography of Morse smooth surfaces.

1. Isolate the brightest points in the image $P = \{p_0, ..., p_{n-1}\}$. Those are the maxima, minima, and saddle points.

2. Let n be a counter of the number of candidate surfaces.

3. Compute $z_i(x, y)$ as a downward process for each point $p_i \in P$, considering p_i to be a maximum point.

4. For each z_i find the point p_j, such that $p_j \neq p_i$, $z_i(p_j)$ is a degenerate (inflection) point, and has the highest $z_i(p_j)$ value among all other inflections at the P locations.

5. Merge $z_i \cap z_k$ that share the same degenerate (inflection) point p_j, and their highest inflection p_j level curves osculate only at p_j. The merge step is

$$z_n(x, y) = \max\{z_i(x, y), z_k(x, y) + (z_i(p_j) - z_k(p_j))\}.$$

Next, detect the highest inflection point for z_n as before and add it to the active candidate list. Increase the counter $n = n + 1$.

6. While there are unmerged couples go to step 5.

7. End.

The inflection identification is done by counting the number of intersections of the surface with a circle hovering above the coordinate plane at the height of the tested point. Two intersections corresponds to an inflection point, zero to a maximum or minimum point, and four intersections to a valid saddle point.

Figure 8.8 presents the reconstruction of a complicated surface from its shading image. In this example there are 11 bright points ($|\nabla z| = 0$) that generate 11 initial surfaces before the merging step starts.

For further details the reader is referred to [100, 196].

8.6   Oblique Light Source

Let us focus on the oblique light source case in which the light source direction is different than that of the viewer. We follow [113] and show first that the intensity location depends on the height itself. This dependency can be resolved by a consistent numerical scheme that selects the smallest $\tilde{z}$ value from the neighboring grid points in the numerical update step. This simple selection makes it possible to solve the oblique light source case by the fast marching method as reported in [113] and presented in Chapter 7.

Recall that the shading image for this Lambertian case is given by

$$I(x, y) = \langle \vec{l}, \vec{\mathcal{N}} \rangle,$$

where $\vec{l} = (l_1, l_2, l_3)$ is the light source direction, and $\vec{\mathcal{N}}$, the unit normal to the surface $z(x, y)$ that we want to reconstruct, is given by

$$\vec{\mathcal{N}} = \frac{(-z_x, -z_y, 1)}{\sqrt{1 + z_x^2 + z_y^2}}. \tag{8.3}$$

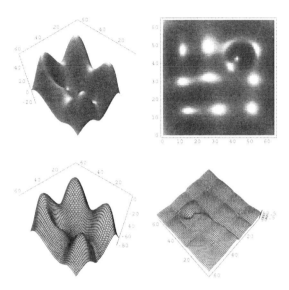

Figure 8.8: A low-resolution 64×64 shading image (top right) of a smooth surface with nontrivial topography (top left), the reconstructed surface (bottom left), and the relative error (bottom right).

We use our freedom to choose the coordinate system so that $l_2 = 0$; this is done by rotating the (x, y) image plane, so that $\vec{l} \cdot (0, 1, 0) = 0$. The shading image is then given by

$$I(x, y) = \langle (l_1, 0, l_3) \cdot \vec{\mathcal{N}} \rangle, \qquad (8.4)$$

where $l_1^2 + l_3^2 = 1$. Equation (8.4) involves the term z_x. It requires some additional thought to construct a monotonic approximation to this term and an appropriate update rule.

If we had had the brightness image in the light source coordinates $\tilde{I}(\tilde{x})$, then the problem would have become the vertical light source case, which is given by the eikonal equation

$$\tilde{z}_{\tilde{x}}^2 + \tilde{z}_y^2 = \frac{1}{\tilde{I}(\tilde{x}, y)^2} - 1; \qquad (8.5)$$

see Figure 8.9.

Lee and Rosenfeld [129] suggest the light source coordinates "to improve" early shape from shading algorithms. In fact, adopting this suggestion, it is simple to view the reflectance map "almost" as an eikonal equation for which we can design a very efficient numerical method. In the light source coordinate system, the equation to solve looks like the eikonal equation, yet the right-hand side depends on the surface itself via

$$\tilde{I}(\tilde{x}, y) = I(l_3 \tilde{x} + l_1 \tilde{z}, y). \qquad (8.6)$$

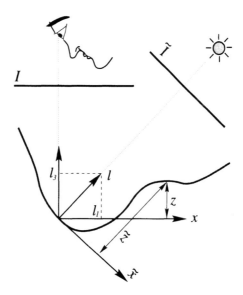

Figure 8.9: In the oblique light source case, the natural coordinate system is determined by the light source [129].

That is, we need to evaluate the value of the surface at a point in order to find the "brightness" and only then plug it to Eq. (8.6) and use the fast marching method to solve Eq. (8.5).

In order to overcome this dependence, we use the fast marching principle and "adopt" the smallest $\tilde{z}$ value from all the neighbors of the updated grid point. The update step for the fast marching method then reads

$$\tilde{z}_1 = \min\{\tilde{z}_{i-1,j}, \tilde{z}_{i+1,j}\}$$
$$\tilde{z}_2 = \min\{\tilde{z}_{i,j-1}, \tilde{z}_{i,j+1}\}$$
$$k = l_3 i + l_1 \min\{\tilde{z}_1, \tilde{z}_2\}$$
if $|\tilde{z}_1 - \tilde{z}_2| < f_{kj}$ then
$$\tilde{z}_{ij} = \frac{\tilde{z}_1 + \tilde{z}_2 + \sqrt{2f_{kj}^2 - (\tilde{z}_1 - \tilde{z}_2)^2}}{2}$$
else $\tilde{z}_{ij} = \min\{\tilde{z}_1, \tilde{z}_2\} + f_{kj}$

where $\tilde{z}_{ij} = z(i\Delta\tilde{x}, j\Delta y)$, and $f_{kj} = f(k\Delta x, j\Delta y)$. Again, without loss of generality we assume $\Delta\tilde{x} = \Delta y = 1$, and $f(x, y) = \sqrt{I(x, y)^{-2} - 1}$. The numerical algorithm in this case is still consistent, one pass is enough since the smallest $\tilde{z}$ neighbor will never change its value, and is thus within the fast marching framework. The map between the light source coordinates $(\tilde{x}, y, \tilde{z})$ and the image coordinates (x, y, z) is a simple rotation given by

$$\begin{pmatrix} x \\ y \\ z \end{pmatrix} = \begin{pmatrix} l_3 & 0 & l_1 \\ 0 & 1 & 0 \\ -l_1 & 0 & l_3 \end{pmatrix} \begin{pmatrix} \tilde{x} \\ y \\ \tilde{z} \end{pmatrix}.$$

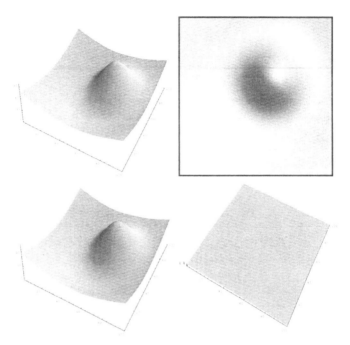

Figure 8.10: Upper left is the original synthetic surface. Upper right its shading image. The reconstruction of the surface from its shading image is shown at the bottom left. The bottom right is the difference between the original surface and its reconstruction.

We have thereby extended the fast marching method to the case of $|\nabla z| = F(z)$ relevant to the oblique light source shape from shading problem. A consistent solution can be computed with $O(N \log N)$, where N is the total number of pixels (grid points).

In [113] the algorithm was tested on a synthetic shading image of the simplest surface with the three basic types of local extremum points: a maximum, a minimum, and a saddle. The oblique light source is given by $\vec{l} = (0.2, 0, 0.96)$. Observe that we do not deal here with self-casting shadows (see [136]), nor with solving the global topological structure (see [100, 62, 24]).

The local extremum points cause singularities at the right-hand side of the eikonal equation since the intensity at their corresponding image locations is equal to zero. This fact should not cause any problem to our numerical algorithm, since one could set the intensity values that are smaller than a given threshold (say $O(\Delta x)$) to some small value (say $O(\Delta x)$) without reducing the global order of accuracy, where Δx is the grid spacing (the distance between two grid points). Figure 8.10 presents the surface, its shading image, the reconstructed surface, and the error, for the oblique light source case. The surface is the solution to Eqs. (8.5) and (8.6) with a fixed value at the minimum point (one of the singular points).

8.7 Summary

We presented an $O(N \log N)$ algorithm for surface reconstruction from its shading image. The computational complexity bound is data independent (unlike other iterative methods [13, 62]). It is the most efficient sequential algorithm for Horn's original formulation of the shape from shading problem and a natural extension and application of the fast marching method.

8.8 Exercises

1. The surface $\mathcal{S}(t)$ evolves by its mean curvature $\partial_t \mathcal{S} = H\vec{\mathcal{N}}$ in time. Given three shading images of the surface $\mathcal{S}(0)$, where each image is taken from the same camera location but with a different light source direction, compute the shading image of the evolving surface without reconstructing the 3D shape of the surface. You may assume the light source directions are given, there are no shadows, and a parallel projection to the image plane. Simulate your result on a synthetic model.

   What are the requirements for differentiability of the surface $\mathcal{S}(t)$, and how would you enforce it along the flow?

2. Research project: Extend the global shape from shading to smooth manifolds. That is, let z be a smooth function defined on the surface $\mathcal{S}$, and assume we have as initial conditions the "image" $I = \nabla_g z$ "painted" on the surface, where ∇_g is the first differential operator of Beltrami, the extension of the gradient magnitude to curved manifolds. The problem is, given $\mathcal{S}$ and I, reconstruct z. Are Morse theory and Morse index enough here? There is a nontrivial extension from singularity theory known as the Conley index.

9

2D and 3D Image Segmentation

An important step toward segmentation of noisy images is edge detection. In many cases we have the luxury to provide the algorithm a good guess of the object boundary. We would like a refinement procedure to act on our coarse guess and adjust it to its "most appropriate" location. For this purpose dynamic models are often used. Deformable curves, active contours, or "snakes" refine a given initial guess of the object boundaries, using local information along the deforming curve. Originally, the motion of "snakes" as introduced by Terzopoulos et al. [90, 204, 203] was determined by an energy functional influenced by external and internal forces. The external forces refer to the way the image gray-levels that are "external" to the curve influence its motion, while internal forces refer to the smoothness and intrinsic elasticity of the curve.

Let us consider an edge penalty function $g(x, y) : \mathbb{R}^2 \to \mathbb{R}^+$, where for example $g(x, y) = (1 + |\nabla I(x, y)|^2)^{-1/2}$, and $I(x, y) : \mathbb{R}^2 \to \mathbb{R}^+$ is the data image. The function g indicates the presence of edges in the image. That is, high gradient magnitudes in the image that indicate the possible presence of an edge are mapped by g to small values, while flat regions in the image are mapped by g to one. The first snakes models minimized an energy functional along a general parametric curve $C(p) : [0, 1] \to \mathbb{R}^2$ given by

$$E(C) = \int_0^1 \left(\alpha |C_p(p)|^2 + \beta |C_{pp}(p)|^2 + g(C(p)) \right) dp.$$

The first two terms, multiplied by the two positive constants α and β, measure the smoothness and "elasticity" of the curve referred to as *internal energy*. The third *external energy* term penalizes the nonedginess along the curve, and should pull the deforming curve toward the high gradients in the image.

The Euler–Lagrange equation for $E(C)$ is given by

$$-\frac{d}{dp}\left(\alpha C_p(p)\right) + \frac{d^2}{dp^2}\left(\beta C_{pp}(p)\right) + \nabla g(C(p)) = 0.$$

The snake model uses these EL equations for evolving an initial given parameterized contour that converges to a local minimum of the functional via the gradient descent flow

$$C_t = \alpha C_{pp} - \beta C_{pppp} - \nabla g(C(p)).$$

This flow is not geometric, which means that for different parameterizations of the same curve we could get different final results. As we will see, it is possible to limit these set of results via the Maupertuis principle of least action, and to link the case where $\beta = 0$ to an intrinsic functional up to a single constant that captures the effect of the initial parameterization. A remedy to the dependency on the parameterization is to modify the energy functional to be geometric.

9.1   The Level Set Geometric Model

Geometric flows for segmentation were introduced by Caselless et al. and by Malladi et al. in [33, 144, 145, 146]. In both cases the idea is to use the level set formulation of a geometric evolution equation to handle topological changes of the curve and overcome numerical difficulties. The basic model includes a constant external motion coupled with a geometric smoothing term. The curve evolution is

$$C_t = (\kappa - v)g(C(p))\vec{N}$$

where v is a constant, and g is an edge indicator stopping function that should stop the evolution as the contour approaches a desired configuration. The level set formulation for this evolution is

$$\phi_t = \left(\mathrm{div}\left(\frac{\nabla \phi}{|\nabla \phi|}\right) + v\right)\tilde{g}(x,y)|\nabla \phi|,$$

where $\tilde{g}(x,y)$ is an extension of the $g(x,y)$ values along the zero set to the rest of the domain. Although geometric, the above flow is not a minimization of a meaningful functional, and this segmentation model still needs a fix.

9.2   Geodesic Active Contours

A unification of both the parametric and the level sets geometric models was first introduced by Caselles et al. [34, 35], by a model known as the *geodesic active contours*; see [91, 193] for a related effort. The functional is now defined geometrically as

$$\int g(C(s))ds,$$

where s is the Euclidean arclength, for which the steepest descent flow is

$$C_t = (\kappa g - \langle \nabla g, \vec{N} \rangle)\vec{N}$$

(see Figure 9.1), with the level set formulation

$$\phi_t = g\mathrm{div}\left(\frac{\nabla\phi}{|\nabla\phi|}\right)|\nabla\phi| + \langle \nabla g, \nabla\phi \rangle,$$

or in a more compact way

$$\phi_t = \mathrm{div}\left(g\frac{\nabla\phi}{|\nabla\phi|}\right)|\nabla\phi|.$$

Given an initial contour, this flow deforms the curve toward the path of minimal weighted length, where the arclength is now measured by $g^2 ds^2 = g^2(x,y)(dx^2 + dy^2)$. Note that this is exactly the measure by which light rays pick the shortest paths in an isotropic nonhomogeneous interface. This is the distance measure defined by Fermat that leads to Snell's law in optics.

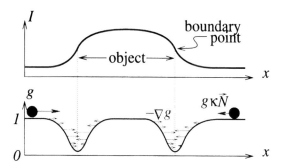

Figure 9.1: A 1D slice shows the geodesic active contours under the influence of geometric forces: internal $\kappa g\vec{N}$, and external $\langle \nabla g, \vec{N} \rangle\vec{N}$. The two circles represent the two cross-section points as a result of slicing the contour for the 1D presentation.

Practically, in order to accelerate the motion of the active contours, a constant velocity term $v(t)g(x,y)$ that is reduced in time is added. The geodesic active contour flow is then given by

$$\phi_t = \left(\mathrm{div}\left(g(|\nabla I|)\frac{\nabla\phi}{|\nabla\phi|}\right) + g(|\nabla I|)v(t)\right)|\nabla\phi|,$$

where, for example, $v(t) = 1/(1 + t^2)$. In some cases we prefer to have an expanding contour, in which the constant flow v is applied with a minus sign. It would then act like a "balloon force," first introduced and used by Cohen in [46, 45].

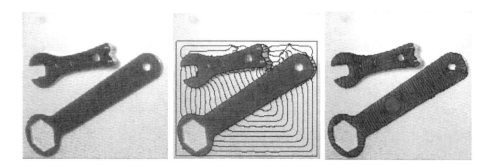

Figure 9.2: Segmentation via the geodesic active contour by inwards and outwards propagation.

Figure 9.2 shows a geodesic active contour propagating inwards from an initial rectangular frame and outwards from two initial circles inside the objects, and eventually locking at the boundaries of the objects. Figure 9.3 demonstrates segmentation of a fetus from a noisy ultrasound image. The initialization is a union of four disks.

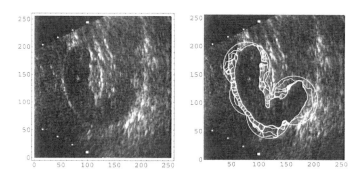

Figure 9.3: Segmentation of a fetus in an ultrasound image.

9.3   Relation to Image Enhancement Methods

Recall that for $\int F(\phi_x)dx$ the EL equation is given by $(d/dx)\partial_{\phi_x}F = 0$, and for $\int G(\sqrt{\phi_x^2 + \phi_y^2})dxdy$ we have the EL $\text{div}(G'(|\nabla\phi|)\nabla\phi/|\nabla\phi|) = 0$.

The geodesic active contours search for a path along which the integration $\int g(C(s))ds$ is minimal, while the above EL is an extremum condition for $\phi(x, y)$, which is a 2D graph surface rather than a 1D curve.

More specifically, given $G(|\nabla\phi|)$, such that $g(v) = G'(v)$, we have that the EL of

$$\int G(|\nabla\phi|)dxdy$$

is given by

$$\text{div}\left(g(|\nabla\phi|)\frac{\nabla\phi}{|\nabla\phi|}\right) = 0.$$

The steepest descent minimization flow in this case is $\phi_t = \text{div}(g(|\nabla\phi|)\frac{\nabla\phi}{|\nabla\phi|})$. It can be written through its level set contours evolution

$$C_t = \frac{1}{|\nabla\phi|}(\kappa g(|\nabla\phi|) - \langle\nabla g(|\nabla\phi|), \vec{N}\rangle)\vec{N}.$$

The geodesic active contours fix g as a function of the original image throughout the flow. However, other models allow g to dynamically change. One interesting example is $G(v) = v$, which is the total variation method (TV method), introduced by Osher, Rudin, and Fatemi [173] as good measure for gray-level image denoising.

Other variations on the geodesic active contours are possible. For example, we can add the area Ω of the interior of the contour, $\partial\Omega = C$, as a penalty to our geometric energy, which now reads

$$E(C) = \int_C g(C)ds + \alpha\mathcal{A}(\Omega) = \int_C g(C)ds + \alpha\iint_\Omega da.$$

In this case, the EL equations are

$$(\kappa g - \langle\nabla g, \vec{N}\rangle - \alpha)\vec{N} = 0.$$

Using the freedom of parameterization in the steepest descent flow and multiplying the evolution by g, we obtain the flow

$$C_t = (g^2\kappa - g\langle\nabla g, \vec{N}\rangle - \alpha g)\vec{N}.$$

9.4   Nongeometric Measures and the Maupertuis Principle of Least Action

The Maupertuis principle of least action deals with actions of the form

$$\int F(x, y, x_p, y_p, p)dp,$$

that are free of the parameterization p, and thus may be written as

$$\int F(x, y, x_p, y_p)dp.$$

The Maupertuis principle of least action states that the extremum curves of such functionals are equivalent to the extremum curves of functionals that are independent of p. More specifically, for our case of geodesic active contours, the dual functional is actually a geometric one, or intrinsic. This intrinsic functional (also known as geometric, or reparameterization invariant) can be defined as a function of the original measure, $F()$, and an arbitrary scalar constant. That is, up to a constant, we can find a geometric measure for which the extremum curves are the same as those of the nongeometric one. Here we first deal with the general functional, and then give the geometric connection between the geodesic active contour and classical snakes as a specific example.

Proof. Consider the functional

$$\int F(C, C_p)dp.$$

By the chain rule

$$\frac{dF}{dp} = \langle C_{pp}, \{F_{x_p}, F_{y_p}\}\rangle + \langle \nabla F, C_p\rangle,$$

where $\nabla F = \{F_x, F_y\}$. Recall that the EL equation for the functional is given by

$$\nabla F = \frac{d}{dp}\{F_{x_p}, F_{y_p}\}.$$

We plug the EL into the $(d/dp)F$ equation and obtain

$$\begin{aligned}
\frac{d}{dp}F &= \langle C_{pp}, \{F_{x_p}, F_{y_p}\}\rangle + \langle C_p, \frac{d}{dp}\{F_{x_p}, F_{y_p}\}\rangle \\
&= \frac{d}{dp}\langle C_p, \{F_{x_p}, F_{y_p}\}\rangle.
\end{aligned}$$

We can now write

$$\frac{d}{dp}\left(F - \langle C_p, \{F_{x_p}, F_{y_p}\}\rangle\right) = 0,$$

and integration over p yields

$$F - \langle C_p, \{F_{x_p}, F_{y_p}\}\rangle = \text{constant}. \tag{9.1}$$

Let us search for all extremum curves of the functional with the additional constant $-W$, which should not change the result. Now, we change the parameterization from p to x, such that $y_p = y_x x_p$, and x_p can be extracted from Eq. (9.1), say

$$x_p = \Psi(y_x, x, y, W).$$

Then, by subtracting the constant W we can write a new functional for which the extremum curves are the same as our original one,

$$
\begin{aligned}
\int_{p_0}^{p_1} (F - W)dp &= \int_{p_0}^{p_1} \langle C_p, \{F_{x_p}, F_{y_p}\}\rangle dp \\
&= \int_{p_0}^{p_1} (F_{x_p} x_p + F_{y_p} y_p)dp \\
&= \int_{x_0}^{x_1} (F_{x_p} + y_x F_{y_p})dx \\
&= \int_{x_0}^{x_1} f(y_x, x, y, W)dx. \quad\blacksquare
\end{aligned}
$$

This new functional is independent of p, yet, in its general form it is not reparameterization-invariant and may depend on the specific choice of the xy-coordinates. We are interested in more specific measures, those that are rotation- and translation-invariant. An important example:
Consider the functional $\int_0^1 (g(x,y) + |C_p|^2)dp$, where $F = g(x,y) + x_p^2 + y_p^2$; we have

$$W = F - \langle C_p, \{F_{x_p}, F_{y_p}\}\rangle = g + (x_p^2 + y_p^2) - 2x_p^2 - 2y_p^2;$$

thus $|C_p|^2 = g - W$. Now, changing of parameterization from p to x we have

$$(x_p^2 + y_x^2 x_p^2) = g - W,$$

which yields

$$x_p = \sqrt{\frac{g - W}{1 + y_x^2}}.$$

Now, the minimization of the functional

$$
\begin{aligned}
\int_{x_0}^{x_1} (F_{x_p} + y_x F_{y_p})dx &= \int_{x_0}^{x_1} (2x_p + y_x 2y_p)dx \\
&= 2\int_{x_0}^{x_1} \sqrt{\frac{g - W}{1 + y_x^2}}(1 + y_x^2)dx \\
&= 2\int_{x_0}^{x_1} \sqrt{g - W}\sqrt{1 + y_x^2}dx
\end{aligned}
$$

is reparameterization-invariant, and therefore identical to the minimization of the geometric functional $\int \sqrt{g - W} ds$, where s is the Euclidean arclength.

Since the parameterization is arbitrary to begin with, the final solution would be the result of minimizing the geometric functional $\int \sqrt{g - W} ds$ with an arbitrary constant W. Introduction of an arbitrary parameter to the segmentation procedure is usually undesired. If one insists on using a linear model that results in a nongeometric functional, one should be aware that different initial conditions such as different parameterization of the same initial contour would yield results that correspond to the minimization of different geometric functionals.

Note that for the geometric functionals we have a unique constant $W = 0$. For example, for $F = g(C)|C_p|$ we have

$$g|C_p| - gx_p \frac{x_p}{|C_p|} - gy_p \frac{y_p}{|C_p|} = 0.$$

9.5   Edge Integration

In [47] an efficient direct edge integration method that minimizes the geodesic active contour functional was introduced. The idea is to replace the variational method with a direct approach that finds the shortest weighted path connecting two points along the boundary. The steps are as follows: We first show that the geodesic active contours at steady state are the flow lines of a viscosity solution T of an eikonal equation. The surface solution $T : \Omega \subset \mathbb{R}^2 \to \mathbb{R}$ can be efficiently computed by the fast marching method on a rectangular grid. Next, we use a simple back-tracking technique to integrate the optimal path along the steepest descent curves, or "flow lines," of T between the pixel grid points.

Let T be a weighted distance function, defined such that

$$|\nabla T| = g(x, y),$$

with given boundary conditions $T(x_0, y_0) = 0$. Let us first show that the "flow lines" of T defined by the ODE

$$\frac{dC}{dp} = \nabla T$$

satisfy the EL equation of the geometric measure

$$\int g(C)|C_p|dp = \int g(C)ds.$$

Lemma 8 *The flow line curves defined by $C_p = \nabla T$, where $|\nabla T| = g$, satisfy the EL of $\int g(C)ds$, where s is the Euclidean arclength.*

Proof. The EL equation is given by $(\kappa g - \langle \nabla g, \vec{N} \rangle)\vec{N} = 0$. By definition we have

$$\kappa\vec{N} \equiv C_{ss} = \frac{\partial}{\partial s}\left(\frac{\nabla T}{|\nabla T|}\right)$$

$$= \left\{\left\langle \nabla\left(\frac{T_x}{|\nabla T|}\right), C_s \right\rangle, \left\langle \nabla\left(\frac{T_y}{|\nabla T|}\right), C_s \right\rangle\right\}$$

$$= \left\{\left\langle \nabla\left(\frac{T_x}{|\nabla T|}\right), \frac{\nabla T}{|\nabla T|} \right\rangle, \left\langle \nabla\left(\frac{T_y}{|\nabla T|}\right), \frac{\nabla T}{|\nabla T|} \right\rangle\right\}.$$

Using the relations $g = |\nabla T|$, $\nabla g = \nabla(|\nabla T|)$, and $\vec{N} = \{T_y, -T_x\}/|\nabla T|$, we obtain

$$\kappa\vec{N} = \frac{\langle \nabla g, \vec{N} \rangle}{g}\vec{N},$$

which concludes the proof. ∎

The practical meaning of this relation is that we can use the efficient and accurate fast marching method that was developed for the eikonal case to compute the weighted distance function T from a given source point (x_0, y_0) along the boundary. Next, given T, and a second point (x_1, y_1), we need to solve the ODE $C_p = -\nabla T$, with initial condition $C(0) = (x_1, y_1)$ and obtain the desired edge.

Figure 9.4 shows tracking vessels in a medical angiographic image of the eye fundus.

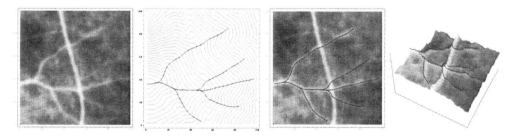

Figure 9.4: Tracking vessels as multiple path detection to a start point in an angiographic image: Left to right: Original image, the paths drawn on the level sets of $T(x, y)$, on the intensity image, and on the intensity image as a height map. In this case $g(x, y) = 1/I(x, y)$.

9.6   Geometric Segmentation in 3D

In a very similar way to the two-dimensional case, we consider the parametric surface $\mathcal{S}(u,v) : \mathbb{R}^2 \to \mathbb{R}^3$, and an edge indicator function $g : \mathbb{R}^3 \to \mathbb{R}^+$, for example,

$$g(x,y,z) = \frac{1}{1 + |\nabla I|^2} = \frac{1}{1 + I_x^2 + I_y^2 + I_z^2},$$

where $I : \mathbb{R}^3 \to \mathbb{R}^+$ is a 3D intensity data image. The functional to minimize now reads

$$
\begin{aligned}
E(\mathcal{S}) &= \iint g(\mathcal{S})da \\
&= \iint g(\mathcal{S}(u,v))|\mathcal{S}_u \times \mathcal{S}_v|dudv.
\end{aligned}
$$

The Euler–Lagrange equations are given by

$$gH\vec{N} - \langle \nabla g, \vec{N}\rangle \vec{N} = 0,$$

where H is the mean curvature of $\mathcal{S}$. The steepest descent flow now reads

$$\mathcal{S}_t = gH\vec{N} - \langle \nabla g, \vec{N}\rangle \vec{N},$$

or in its implicit level set form, we have as before

$$\phi_t = \text{div}\left(g(|\nabla I|)\frac{\nabla \phi}{|\nabla \phi|}\right)|\nabla \phi|, \tag{9.2}$$

which was named minimal segmentation active surfaces in [37].

ϕ is now defined on the 3D space, $\phi : \mathbb{R}^3 \times [0, \tau) \to \mathbb{R}^+$. The marching cube algorithm is a useful rendering technique of the relevant level set, which is sometimes referred to as iso-surface rendering. See Figure 9.5, taken from [37], for an example of topology change in 3D, where an initial ellipsoid changes its topology into two linked tori.

Figure 9.5: Segmentation of two tori.

Evolving the whole volume is, in general, computationally expensive. There are several acceleration methods. One example is an efficient initialization based on the fast marching method [143]. The AOS (additive operator splitting) method, which is an unconditionally stable numerical technique, was introduced in [137, 138] and first applied to image enhancement and symmetric scale space in [212, 213]. The numerical method was coupled with an efficient distance map algorithm in [80] for accelerating the geodesic active contours. An alternative multiplicative scheme was later used in [96]. The narrow band approach [43, 44, 1] limits the calculations to a neighborhood of the zero set, while pyramidal techniques [162] solve the problem at a coarse resolution and interpolate the rough solution as an initial condition at the finer grid. An extension of the geometric segmentation was found to be useful in 3D shape reconstruction from several images, also known as the shape from stereo problem, first in [69], and later in [95]; see Figure 9.6.

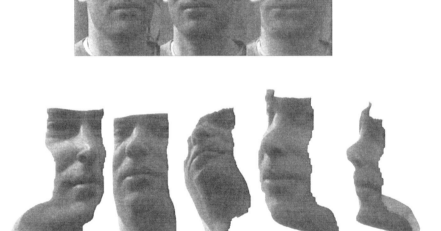

Figure 9.6: Shape from stereo: Top: The stereo pair and their normalized addition. Bottom: Reconstructed surface after weighted area minimization.

9.7 Efficient Numerical Schemes

We follow [96] and describe efficient methods for numerically implementing geometric active contour-like segmentation procedures. In [213] Weickert et al. use an unconditionally stable, and thus efficient, numerical scheme for nonlinear isotropic image diffusion known as *additive operator splitting* (AOS), that was first introduced in [137, 138], and has some nice symmetry and parallel properties. Goldenberg et al. [79] couple the AOS with Sethian's fast marching on regular grids [190] (see [205, 41] for related approaches), with multiresolution [163], and with the narrow band approach [42, 1], as part of their fast geodesic active contour model for segmentation and object tracking in video sequences. Motivated by these results, we can extend the efficient numerical methods for the geodesic active contour [35] presented in [79], for the variational edge integration models introduced in [102, 103], and for the minimal variance [39]. In this section we review efficient numerical schemes and modify them in order to solve the level set formulation of edge integration and object segmentation problem in image analysis.

Let us analyze the following level set formulation,

$$\phi_t = \left(\alpha \operatorname{div} \left(g(x,y) \frac{\nabla \phi}{|\nabla \phi|} \right) + \eta(\phi, \nabla I) \right) |\nabla \phi|,$$

$$\eta(\phi, \nabla I) = \operatorname{sign}(\langle \nabla I, \nabla \phi \rangle) \Delta I + \beta(c_2 - c_1) \left(I - \frac{c_1 + c_2}{2} \right).$$

If we assume $\phi(x, y; t)$ to be a distance function of its zero set, then we could approximate the short time evolution of the above equation by setting $|\nabla \phi| = 1$. The short time evolution for a distance function ϕ is thereby

$$\phi_t = \alpha \operatorname{div} \left(g(x,y) \nabla \phi \right) + \eta(\phi, \nabla I)$$
$$= \alpha \frac{\partial}{\partial x} \left(g(x,y) \frac{\partial}{\partial x} \phi \right) + \alpha \frac{\partial}{\partial y} \left(g(x,y) \frac{\partial}{\partial y} \phi \right) + \eta(\phi, \nabla I).$$

Note, that when using a narrow band around the zero set to reduce computational complexity on sequential computers, the distance from the zero set needs to be recomputed in order to determine the width of the band at each iteration. Thus, there is no additional computational cost in simplifying the model while considering a distance map rather than an arbitrary smooth function. We thereby enjoy both the efficiency of the simplified and almost linear model, and the low computational cost of the narrow band.

Denote the operators

$$\mathbf{A}_1 = \frac{\partial}{\partial x} g(x,y) \frac{\partial}{\partial x},$$
$$\mathbf{A}_2 = \frac{\partial}{\partial y} g(x,y) \frac{\partial}{\partial y}.$$

Using these notations we can write the evolution equation as

$$\phi_t = \alpha \left(\mathbf{A}_1 + \mathbf{A}_2 \right) \phi + \eta(\phi, \nabla I).$$

Next, we approximate the time derivative using forward approximation $\phi_t \approx (\phi^{n+1} - \phi^n)/\tau$, which yields the explicit scheme

$$\phi^{n+1} = \phi^n + \tau \left(\alpha \left(\mathbf{A}_1 + \mathbf{A}_2 \right) \phi^n + \tau \eta(\phi^n, \nabla I) \right)$$

$$= \left(\mathbf{I} + \tau\alpha(\mathbf{A}_1 + \mathbf{A}_2)\right)\phi^n + \tau\eta(\phi^n, \nabla I),$$

where, after sampling the x, y-plane, $\mathbf{I}$ is the identity matrix and I is our input image. The operators $\mathbf{A}_l$ become matrix operators, and ϕ^n is represented as a vector in either column or row stack, depending on the acting operator. This way, the operators $\mathbf{A}_l$ are tridiagonal, which makes its inverse computation fairly simple using the Thomas algorithm. Note that in the explicit scheme there is no need to invert any operator, yet the numerical time step is bounded for stability.

Let us first follow [213], and use a simple discretization for the $\mathbf{A}_l$, $l \in \{1, 2\}$ operators. For a given row, let

$$\frac{\partial}{\partial x}\left(g(x)\frac{\partial}{\partial x}\phi\right) \approx \sum_{j \in \mathcal{N}(i)} \frac{g_j + g_i}{2h^2}\left(\phi_j - \phi_i\right),$$

where $\mathcal{N}(i)$ is the set $\{i-1, i+1\}$, representing the two horizontal neighbors of pixel i, and h is the space between neighboring pixels. The elements of A_1 are thereby given by

$$a_{ij} = \begin{cases} \frac{g_i + g_j}{2h^2}, & j \in \mathcal{N}(i), \\ -\sum_{k \in \mathcal{N}(i)} \frac{g_i + g_k}{2h^2}, & j = i, \\ 0, & \text{else}. \end{cases}$$

In order to construct an unconditionally stable scheme we use a locally one-dimensional (LOD) scheme adopted for our problem. We use the following scheme

$$\phi^{n+1} = \prod_{l=1}^{2}\left(\mathbf{I} - \tau\alpha\mathbf{A}_l\right)^{-1}\left(\phi^n + \tau\eta(\phi^n, \nabla I)\right).$$

In one dimension it is also known as fully implicit, or backward Euler scheme. It is a first-order implicit numerical approximation, since we have that

$$(\mathbf{I} - \tau\mathbf{A}_1)^{-1}(\mathbf{I} - \tau\mathbf{A}_2)^{-1}(\phi + \tau\eta)$$
$$= \left(\mathbf{I} - \tau\mathbf{A}_1 - \tau\mathbf{A}_2 + O(\tau^2)\right)^{-1}(\phi + \tau\eta)$$
$$= \phi + \tau(\mathbf{A}_1 + \mathbf{A}_2)\phi + \tau\eta + O(\tau^2),$$

where we applied the Taylor series expansion in the second equality. First-order accuracy is sufficient, as our goal is the steady-state solution. We also included the weighted region balloon, minimal variance, and the alignment terms within this implicit scheme, while keeping the first-order accuracy and stability properties of the method. The operators $\mathbf{I} - \tau\mathbf{A}_l$ are positive definite symmetric operators, which make the implicit process unconditionally stable, using either the above multiplicative or the additive (AOS) schemes. If we have an indication that the contour is getting closer to its

final destination, we could switch to an explicit scheme for the final refinement steps with a small time step. In this case, the time step should
be within the stability limits of the explicit scheme. One could also use
a multiresolution pyramidal approach, where the coarse grid still captures
the details of the objects we would like to detect.

9.8 Exercises

1. What are the steepest descent flows minimizing the following geometric energy functionals: $\min \int g^2(C)ds$, $\min(\int g(C)ds + \alpha L)$, where L
   is the length of the contour.

2. Show that the Euler–Lagrange equations for the nonintrinsic energy

$$E(C) = \int_0^1 \left(\frac{1}{2}|C_p(p)|^2 + g(C(p)) \right) dp$$

is given by $-(d/dp)(C_p(p)) + \nabla g(C(p)) = 0$. Prove that the final result for a curve propagating according to the above EL as a steepest
descent, with reparameterization into arclength "on the fly," converges into a (geometric) steady state only for the special case where
$g =$constant along the final curve.

3. **a.** Show that the maximal curvature magnitude of the final (in steady
   state) geodesic active contour is bounded by

$$|\kappa| \leq \sup_{x,y} \left(\frac{|\nabla g|}{g} \right).$$

b. Show that the maximal curvature magnitude for the final geodesic
active contour with $g(x,y) = c + \hat{g}(x,y)$, where $\hat{g} \geq 0$ and c is a
positive constant, is bounded by

$$|\kappa| \leq \sup_{x,y} \left(\frac{|\nabla g|}{c} \right).$$

c. Show that the maximal curvature magnitude for the final geodesic
active contour with $g(x,y) = c + d(x,y)$, where c is a positive constant,
and $d(x,y)$ is a distance map to a set of feature points, is bounded
by

$$|\kappa| \leq \frac{1}{c}.$$

4. Apply the geodesic active contours to segment an object in a noisy image. Control the smoothness of your result by adding a constant to your g function as indicated by the previous exercise.

5. Modify the geodesic active contours model by an additional intrinsic area measure that reduces its influence as the contour approaches the edges. Use the functional $\lambda \int g ds + (1 - \lambda) \iint dA$, where for example $\lambda(t) = t/(1+t)$. Compare the steepest descent flow of your model to the flows with **a.** $\lambda(t) = 0$, and **b.** $\lambda(t) = 1$. Implement and compare the models.

6. Modify the geodesic active contours model by an additional intrinsic weighted area measure. Use the functional $\lambda \int g ds + \lambda \iint g(x, y) dA$, where the second integral is taken over the closed contour. Implement and compare to the models with various values of λ. Hint: Use Green's theorem to derive the Euler–Lagrange for the weighted area.

7. What are the Euler–Lagrange conditions for the geometric contour that minimizes the intrinsic measure $\int \left| \langle \vec{N}, \nabla I \rangle \right| ds$, where $\vec{N}$ is the normal of the contour, s is the Euclidean arclength, and $\nabla I(x, y)$ is the image gradient field?
   a. Solve for the case where the integral is taken along a closed contour, and the case of an open contour.
   b. Use your solution to explain the Marr–Hildreth edge detector.
   c. Extend your result to give a variational explanation to the Haralick edge detector given by $I_{\xi\xi} = 0$, where $\vec{\xi} \equiv \nabla I / |\nabla I|$.
   Hint: Use the fact that the Laplacian is rotationally invariant such that $\Delta I = I_{xx} + I_{yy} = I_{\xi\xi} + I_{\eta\eta}$, where $\vec{\eta} \perp \vec{\xi}$ is the level set direction (orthogonal to the gradient).

8. Optional research project: Add a second-order term to your minimization scheme $\int \kappa^2 ds$. This term introduces elastica to the model. Apply the new model to 2D image segmentation. Next, extend the scheme to 3D and try to implement it. Expect numerical problems in implementing the fourth-order derivatives.

   For a curve embedded in a Riemannian manifold, the curvature of an elastica curve is a critical point of $\int (\kappa^2 + \lambda) ds$, which yields

   $$0 = 2\kappa_{ss} + \kappa^3 + 2\kappa G - \lambda \kappa,$$

   where G is the Gaussian curvature of the manifold, and κ is the geodesic curvature, which we denoted as κ_g in the previous chapters. For example, for a curve embedded in the plane, for which $G = 0$, we have $0 = 2\kappa_{ss} + \kappa^3 - \lambda \kappa$.

9. A small research project: Extend the Cohen–Kimmel edge integration method [47] to segmentation of objects painted on triangulated surfaces.

10. A small research project: Extend the Maupertuis principle to surface evolution and find an example that links between the intrinsic and the parametric models. That is, show that the nonintrinsic functional $\iint F(\mathcal{S}, \mathcal{S}_p, \mathcal{S}_q) dp dq$ is related to an intrinsic functional minimization up to some constants, and present an example.

11. Prove that given the vector field $\vec{V}(x, y) = \{u(x, y), v(x, y)\}$, we have the alignment measure

$$E_A(C) = \oint_C \langle \vec{V}, \vec{N} \rangle ds$$

for which the first variation is given by

$$\frac{\delta E(C)}{\delta C} = \operatorname{div}(\vec{V})\vec{N}.$$

12. Prove that the robust alignment term given by

$$E_{AR}(C) = \oint_C |\langle \vec{V}, \vec{N} \rangle| ds$$

yields the first variation

$$\frac{\delta E_{AR}(C)}{\delta C} = \operatorname{sign}(\langle \vec{V}, \vec{N}(s) \rangle)\operatorname{div}(\vec{V})\vec{N}.$$

13. Prove that the weighted region functional

$$E_W(C) = \iint_{\Omega_C} f(x, y) dx dy$$

yields the first variation

$$\frac{\delta E_W(C)}{\delta C} = -f(x, y)\vec{N}.$$

14. Prove that for the geodesic active contour model

$$E_{GAC}(C) = \oint_C g(C(s)) ds,$$

first variation is given by

$$\frac{\delta E_{GAC}(C)}{\delta C} = -(\kappa g - \langle \nabla g, \vec{N} \rangle)\vec{N}.$$

15. Prove that the first variations for the Chan–Vese minimal variance criterion [39],

$$E_{MV}(C, c_1, c_2) = \frac{1}{2} \iint_{\Omega_C} (I - c_1)^2 dx dy$$
$$+ \frac{1}{2} \iint_{\Omega \backslash \Omega_C} (I - c_2)^2 dx dy,$$

are given by

$$\frac{\delta E_{MV}}{\delta C} = (c_2 - c_1) \left(I - \frac{c_1 + c_2}{2} \right) \vec{N},$$
$$\frac{\delta E_{MV}}{\delta c_1} = \iint_{\Omega_C} I dx dy - c_1 \iint_{\Omega_C} dx dy,$$
$$\frac{\delta E_{MV}}{\delta c_2} = \iint_{\Omega \backslash \Omega_C} I dx dy - c_2 \iint_{\Omega \backslash \Omega_C} dx dy.$$

What are the values of c_1 and c_2 at a minimum of E_{MV}?

16. The robust minimal total deviation criterion is given by

$$E_{RMV}(C, c_1, c_2) = \iint_{\Omega_C} |I - c_1| dx dy + \iint_{\Omega \backslash \Omega_C} |I - c_2| dx dy.$$

Prove that its first variation is

$$\frac{\delta E_{RMV}}{\delta C} = (|I - c_1| - |I - c_2|) \vec{N},$$
$$\frac{\delta E_{RMV}}{\delta c_1} = \iint_{\Omega_C} \text{sign}(I - c_1) dx dy,$$
$$\frac{\delta E_{RMV}}{\delta c_2} = \iint_{\Omega \backslash \Omega_C} \text{sign}(I - c_2) dx dy.$$

What are the values of c_1 and c_2 at a minimum of E_{RMV}?

17. Use your results from the previous exercises to:
    a. Derive the Euler–Lagrange equations for the model defined by

$$E(c_1, c_2, C) = \iint_{\Omega_C} (I - c_1)^2 dx dy$$
$$+ \iint_{\Omega \backslash \Omega_C} (I - c_2)^2 dx dy + \int_C g(C(s)) ds$$
$$+ \int_C |\langle \nabla I, \vec{N} \rangle| ds.$$

Optimize for c_1, c_2, and C as a curve evolution process.

b. Use a robust version given by

$$
\begin{aligned}
E(c_1, c_2, C) \quad = \quad & \iint_{\Omega_C} |I - c_1| dxdy + \iint_{\Omega \backslash \Omega_C} |I - c_2| dxdy \\
& + \int_C g(C(s)) ds.
\end{aligned}
$$

Optimize for c_1, c_2, and C as a curve evolution process.

c. Use one of the implicit operator splitting methods (AOS/LOD/ADI) with a narrow band and redistancing of a level set formulation to implement your evolution algorithm.

10

Geometric Framework in Image Processing

When we process images we would like to improve their visual quality. Usually, we try to modify the image taking into consideration known measures while using prior knowledge like an image formation model or the distortion process.

Assume that during the image acquisition a noise is added and we would like to remove this noise. Let us further assume that the noise has a Gaussian distribution with a known variance added to the intensity values at each pixel. That is, instead of viewing the clean image $I_0(x, y)$ we have a noisy image

$$\tilde{I}(x, y) = I_0(x, y) + n(x, y)$$

where the noise n has a Gaussian distribution with zero mean and σ variance, $n(x, y) = N(0, \sigma)$.

If we have enough statistic samples in the image domain Ω, that is, assuming we have enough pixels to identify the statistics of the noise, we can use the variance definition and write

$$\int_\Omega \left(\tilde{I}(x, y) - I_0(x, y) \right)^2 dxdy = \int_\Omega n^2 dxdy \approx \sigma^2,$$

where w.l.o.g. we assume a unit support for the image, $|\Omega| = 1$.

Now, as we modify an image in order to improve it, we would like to keep the modified image at a "distance" less than σ^2 from the noise image. See Figure 10.1.

Several functionals have been proposed over the years as measures for image quality. These functionals are sometimes referred to as image quality norms. Smoothness, on one hand, and sharp discontinuities between different objects, on the other hand, play an important role in our perception of

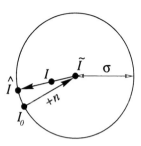

Figure 10.1: The original image is $I_0(x, y)$, an additive noise takes it to $\tilde{I}(x, y) = I_0(x, y) + n(x, y)$. As we denoise the image in order to reconstruct I_0, we would like to keep the modified image at a distance smaller than the noise variance σ from the noisy image.

the image quality. Our first example for such a measure is the L_1 or *total variation* (TV) norm, introduced by Rudin, Osher and Fatemi [173], and defined as

$$TV = \iint_\Omega |\nabla I| dx dy.$$

We would like to minimize the image quality measure TV subject to the constraint

$$\iint_\Omega (I - \tilde{I})^2 dx dy = \sigma^2.$$

There are many ways to minimize a functional subject to constraints; let us explore the one used in [173]. Let $F(I) = |\nabla I|$, and $G(I) = (I - \tilde{I})^2$. We use the EL equation for $\iint_\Omega F dx dy$ as a steepest descent, subject to the constraint. Formally, we write a dynamic process

$$I_t = \frac{\delta F}{\delta I} - \lambda \frac{\delta G}{\delta I}.$$

By simple calculus of variation

$$I_t = \text{div} \left(\frac{\nabla I}{|\nabla I|} \right) - \lambda(I - \tilde{I}).$$

At a steady state we have $I_t = 0$, or

$$\lambda(I - \tilde{I}) = \text{div} \left(\frac{\nabla I}{|\nabla I|} \right).$$

Plugging the last relation to the constraint equation

$$\lambda \sigma^2 = \iint_\Omega \lambda (I - \tilde{I})^2 dx dy$$

$$= \iint_{\Omega} \text{div}\left(\frac{\nabla I}{|\nabla I|}\right)(I - \tilde{I})dxdy.$$

We could have settled for this relation to compute λ, yet, let us further explore this equation

$$\lambda\sigma^2 = \iint_{\Omega}\left[\text{div}\left(\frac{\nabla I}{|\nabla I|}(I - \tilde{I})\right) - \left\langle\frac{\nabla I}{|\nabla I|}, \nabla(I - \tilde{I})\right\rangle\right]dxdy.$$

Using Green's formula

$$\lambda\sigma^2 = \oint_{\partial\Omega}\frac{(I - \tilde{I})}{|\nabla I|}(I_x dy + I_y dx) - \iint_{\Omega}\left\langle\frac{\nabla I}{|\nabla I|}, \nabla I - \nabla\tilde{I}\right\rangle dxdy$$

$$= -\iint_{\Omega}\left(|\nabla I| - \left\langle\frac{\nabla I}{|\nabla I|}, \nabla\tilde{I}\right\rangle\right)dxdy,$$

where we set "natural" boundary conditions to eliminate the integral along the boundary. We conclude with

$$\lambda = -\frac{1}{\sigma^2}\iint_{\Omega}\left(|\nabla I| - \left\langle\frac{\nabla I}{|\nabla I|}, \nabla\tilde{I}\right\rangle\right)dxdy.$$

The final TV reconstruction scheme is given by

$$I_t = \text{div}\left(\frac{\nabla I}{|\nabla I|}\right) - \lambda(I - \tilde{I})$$

$$\lambda = -\frac{1}{\sigma^2}\iint_{\Omega}\left(\sqrt{I_x^2 + I_y^2} - \frac{\tilde{I}_x I_x + \tilde{I}_y I_y}{\sqrt{I_x^2 + I_y^2}}\right)dxdy.$$

Selecting an appropriate smoothness measure for images is subject to empirical tests of human vision and perception, and physical models of image formation. Currently, there is no conclusive measure that captures our general perception of image quality. The above example used the TV or L_1 norm. A different example is the L_2 norm that measures $\iint|\nabla I|^2 dxdy$ and results in the linear heat equation as its steepest descent by the EL equation. Other methods incorporate the constraints within the definition of the norm [14]. The TV norm has several nice qualities and performs better than the L_2 norm in some cases. It can be interpreted geometrically as a selective smoothing of the gray-level sets of the image via the curve evolution equation

$$C_t = \frac{1}{|\nabla I|}\kappa\vec{N},$$

where $1/|\nabla I| = (edge\ indicator)^{-1}$. It is a curve shortening flow in a selective manner: Along the edges the smoothing is suppressed, while at flat regions it acts in a more rapid way.

Note that some regularization should be applied, since one cannot divide by zero, while $|\nabla I|$ can be zero for general images, especially as the image is getting smoother. Some attempts were made to extend the TV and other schemes to the multichannel (color) case. The remainder of this chapter explores the *geometric framework for image processing* that was recently introduced in [198, 118, 119, 107, 108]. It gives links between classical norms and presents a novel solution to the multichannel case.

10.1   Images as Surfaces

Texture and color play important roles in the understanding of many images, specially those that involve natural scenes. It became an important research subject in the fields of psychophysics and computer vision. The study of texture and color perception starts from the pre-image that describes the physics and optics that transform the 3D world into an image. This study tracks human perception from the image formation on the retina and its interpretation at the first perception steps in the brain.

Motivated by human perception and image formation models, the geometric framework introduces a set of tools for enhancing images while preserving either the multichannel edges or the orientation-dependent texture features in them. The images are treated as manifolds in a feature space. This geometrical interpretation led to a simple way for gray-level, color, movies, volumetric medical data, and color-texture image processing. We show how a geometric flow, that is based on the image surface area minimization, yields a proper enhancement procedure for color images. In order to enhance color and texture images we sometimes need to introduce an unstable procedure like the inverse heat equation. We show how to use the geometric framework for this purpose, and explore an extension of Gabor's geometric image sharpening procedure [73] to color images.

The geometric approach views images as a manifold (surface in the 2D case) embedded in a higher dimensional space-feature manifold. In the first example we consider a color image as a two-dimensional surface $\{x, y, R(x, y), G(x, y), B(x, y)\}$ embedded in a five-dimensional space (x, y, R, G, B). As a second example, a volumetric medical image is considered as a three-dimensional manifold $\{x, y, z, I(x, y, z)\}$ embedded in a four-dimensional space (x, y, z, I). Next, a nonlinear scale-space equation is applied to the image. It is derived as a gradient descent of a norm functional that weighs embedding maps in a geometrical way.

The explicit form of the scale-space PDE (or the coupled PDEs) depends on the choice of coordinates and the geometry of the image manifold. In this chapter we limit our discussion to the Euclidean space-feature manifold. One can also choose a non-Euclidean embedding space that introduces *Levi Civita connection* terms as part of the minimization as shown in [197].

We would like to build our iterative minimization schemes such that homogeneous regions are rapidly flattened and simplified while edges are preserved along the scale. An important question, for which there are only partial answers, is how to treat multivalued images in a geometric way. A color image is a good example since one actually considers three images red, green, and blue that are composed into one. The geometric framework attempts to answer this question. An edge-preserving enhancement procedure is a result of minimizing an area norm with respect to the feature coordinates with the natural induced metric, and is expressed via a geometric flow that is named the *Beltrami flow* for image processing.

A popular method for texture analysis is to decompose a given image into a set of subband images using the 2D Gabor–Morlet wavelet transform. Some nice mathematical properties and the relation of this transform to the physiological behavior were studied in [131, 167]. This model was later used for the segmentation, interpretation, and analysis of texture [19, 132], and for texture-based browsing [147]. The Gabor–Morlet wavelet transform was used in [117] to split a given texture image into a set of subband images. Then, the enhancement is performed by a flow in transformed space, that is, the transform coefficients are treated as higher-dimensional manifolds. Other flows in similar feature spaces were proposed in [175, 171, 38, 179, 215]; see also [211] for orientation-preserving flows. These flows are usually based on a flat metric.

The geometric perspective of a color image as a surface embedded in a higher-dimensional space enables us to define a proper coupling in the multichannel color space. Other schemes have also considered image as a surface [14, 65, 218, 142]; some used the image information to build a Riemannian metric for segmentation [34] as we saw in the previous chapter. However, these methods were not generalized to feature space or co-dimensions higher than one.

The geometric framework has the following properties: (1) It is a general way of writing the geometrical scale-space and enhancement algorithms for gray-scale, color, volumetric, time-varying, and texture images; (2) it unifies many existing partial differential equation-based schemes for image processing; (3) it leads to schemes that are feature-preserving and hence suitable for enhancement and segmentation tasks; and (4) it offers a proper coupling between channels in a multichannel image processing.

10.2   The Geometric Framework

Let us first review the geometrical framework in which images are considered as surfaces. Suppose we have an n-dimensional manifold X with coordinates $x^1, x^2, \ldots, x^n$ embedded in an m-dimensional manifold Y with coordinates $y^1, y^2, \ldots, y^m$, where $m > n$. The embedding map $E : X \to Y$

is given explicitly by the m functions of n variables

$$E : (x^1, \dots, x^n) \to \{y^1(x^1, \dots, x^n), \dots, y^m(x^1, \dots, x^n)\}.$$

If we denote the image plane coordinates by x^1 and x^2, then a possible embedding map of a gray-level image is

$$\{y^1(x^1, x^2) = x^1, y^2(x^1, x^2) = x^2, y^3(x^1, x^2) = I(x^1, x^2)\},$$

where $I(x^1, x^2)$ is the intensity. If we further denote $y^1 \equiv x$ and $y^2 \equiv y$ then it can be written with a slight abuse of notations as $\{x, y, I(x, y)\}$.

In order to consider the geometry of the manifold we introduce Riemannian structure, namely a metric. The metric at a given point on the manifold describes the way we measure distances without being dependent on the coordinates, that is, the metric on X locally measures the distances at a point as follows:

$$ds^2 = g_{\mu\nu} dx^\mu dx^\nu, \qquad \mu, \nu \in \{1, \dots, n\},$$

and summation is implied on identical indices. Similarly on Y

$$ds^2 = h_{ij} dy^i dy^j, \qquad i, j \in \{1, \dots, m\}.$$

In an embedding that preserves length these two line elements are equal. Applying the chain rule $dy^i = \partial_\mu y^i dx^\mu$, where $\partial_\mu \equiv \partial/\partial x^\mu$, and summing over μ, yields the induced metric formula

$$g_{\mu\nu} = h_{ij} \partial_\mu y^i \partial_\nu y^j.$$

For the embedding of a gray-level image in an Euclidean three-dimensional space, $h_{ij} = \delta_{ij}$, we obtain the following metric:

$$g_{\mu\nu} = \begin{pmatrix} 1 + I_x^2 & I_x I_y \\ I_x I_y & 1 + I_y^2 \end{pmatrix},$$

where $x^1 \equiv x$ and $x^2 \equiv y$.

Denote by (X, g) the image manifold and its metric and by (Y, h) the space-feature manifold and its metric, then the map $E : X \to Y$ has the following weight:

$$S[y^i, g_{\mu\nu}, h_{ij}] = \int d^n x \sqrt{g} g^{\mu\nu} \partial_\mu y^i \partial_\nu y^j h_{ij}(\mathbf{E}), \qquad (10.1)$$

where n is the dimension of X, g is the determinant of the image metric, $g^{\mu\nu}$ is the inverse of the image metric, the range of indices is $\mu, \nu = 1, \dots, \dim X$, and $i, j = 1, \dots, \dim Y$, and h_{ij} is the metric of the embedding space. For more details see [198]. This is a generalization of the L_2 norm to manifolds.

Many scale-space methods, linear and nonlinear, can be shown to be gradient descent flows of this functional with an appropriately chosen metric of the image manifold. The gradient descent equation is

$$y_t^i = -\frac{1}{\sqrt{g}}\frac{\delta S}{\delta y^i}.$$

The metric is a free parameter of the framework and different choices lead to different scale-space schemes as shown in [197]. For the natural choice of the metric as the induced metric the norm simply becomes the area or the volume of the image manifold, and the flow is toward a minimal surface solution. Minimizing the area action with respect to the feature coordinate (fixing the x- and y-coordinates), we obtain the area minimization direction given by applying the *second-order differential operator of Beltrami* on the feature coordinates. Filtering the image based on this result yields an efficient geometric flow for smoothing the image while preserving the edges. It is written as

$$\mathbf{I}_t = \Delta_g \mathbf{I}, \tag{10.2}$$

where for color $\mathbf{I} = (R, G, B)$. The Beltrami operator, denoted by Δ_g, that is acting on $\mathbf{I}$ is a generalization of the Laplacian from flat spaces. It is defined by [1]

$$\Delta_g \mathbf{I} \equiv \frac{1}{\sqrt{g}}\partial_\mu(\sqrt{g}g^{\mu\nu}\partial_\nu \mathbf{I}). \tag{10.3}$$

For gray-level or color 2D images, the flow is given by

$$I_t^i = \frac{1}{g}\left(p_x^i + q_y^i\right) - \frac{1}{2g^2}\left(g_x p^i + g_y q^i\right), \tag{10.4}$$

where $g_x = \partial_x g \ (g_y = \partial_y g)$, $g_{\mu\nu} = \delta_{\mu\nu} + \sum_i I_\mu^i I_\nu^i$, $g = g_{11}g_{22} - g_{12}^2$, and

$$p^i = g_{22}I_x^i - g_{12}I_y^i, \qquad \text{and} \qquad q^i = -g_{12}I_x^i + g_{11}I_y^i. \tag{10.5}$$

See Figure 10.2 for a matlab code of the Beltrami flow in color R, G, B log domain, where $ds^2 = dx^2 + dy^2 + (d\log R)^2 + (d\log G)^2 + (d\log B)^2$.

For the gray-level case, the above evolution equation is the mean curvature flow of the image surface divided by the induced metric $g = \det(g_{\mu\nu})$. It is the evolution via the $\mathbf{I}$ components of the mean curvature vector $\mathbf{H}$. That is, for the surface $(\mathbf{x}(x^1, x^2), \mathbf{I}(x^1, x^2))$ in the Euclidean space $(\mathbf{x}, \mathbf{I})$, the curvature vector is given by $\mathbf{H} = \Delta_g(\mathbf{x}(x^1, x^2), \mathbf{I}(x^1, x^2))$. If we identify $\mathbf{x}$ with X, then $\Delta_g I^i(\mathbf{x}) = \mathbf{H} \cdot \hat{I}^i$, where this direct computation applies for co-dimensions > 1. The determinant of the induced metric matrix

---

[1]In case the embedding space is chosen non-Euclidean there is an extra term. See [198].

```
%%%%%%%%%%%%%%%%%%%%%%%%%%%%%%%%%%%%%%%%%%
function O = beltrami(tif_image,iterations)
    A=imread(tif_image);
    O=log2(double(A)+1);          % work in the log domain
    R = double(O(:,:,1)); G = double(O(:,:,2)); B= double(O(:,:,3));
    beta = 0.01; dt = 0.21*beta;
    for p=1:iterations,
        Rx = Dmx(R); Gx = Dmx(G); Bx = Dmx(B);
        Ry = Dmy(R); Gy = Dmy(G); By = Dmy(B);

        g11 = beta+ Rx.^2+Gx.^2+Bx.^2;
        g12 = Rx.*Ry+Gx.*Gy+Bx.*By;
        g22 = beta+ Ry.^2+Gy.^2+By.^2;
        gm05 = (g11.* g22 - g12.^2).^(-0.5);       % gm05 = 1/√g

        R = belt(dt,R,Rx,Ry,gm05,g11,g12,g22);
        G = belt(dt,G,Gx,Gy,gm05,g11,g12,g22);
        B = belt(dt,B,Bx,By,gm05,g11,g12,g22);
    end;
    O(:,:,1)=(2.^R) -1;O(:,:,2)=(2.^G) -1;O(:,:,3)=(2.^B) -1;
%%%%%%%%%%%%%%%%%%%%%%%%%%%%%%%%%%%%%%%%%%
% res = R +dt Δ_g R
function res = belt(dt,R,Rx,Ry,gm05,g11,g12,g22)
    res = gm05.* ...
    ( Dpx(gm05.*( g22.*Rx-g12.*Ry))+Dpy(gm05.*(-g12.*Rx+g11.*Ry)) );
    res = R + dt* res; res = max(min(res,8),0);
%%%%%%%%%%%%%%%%%%%%%%%%%%%%%% Backwards derivatives
function f = Dmx(P)
    f = P - P([1 1:end-1],:);
%%%%%%%%%%%%%%%%%%%%%%%%%%%%%%%%%%%%%%%%%%%%
function f = Dmy(P)
    f = (Dmx(P'))';
%%%%%%%%%%%%%%%%%%%%%%%%%%%%%%%% Forward derivatives
function f = Dpx(P)
    f = P([2:end end],:) - P;
%%%%%%%%%%%%%%%%%%%%%%%%%%%%%%%%%%%%%%%%%%%%%
function f = Dpy(P)
    f = (Dpx(P'))';
```

Figure 10.2: Matlab procedure for the Beltrami flow in the log domain.
The effect is apparent after a few iterations. Run for example: O = beltrami('image.tif',20);

$g = \det(g_{\mu\nu})$ may be considered as a generalized form of an edge indicator. Therefore, the flow defined by Eq. (10.2) is a selective smoothing mechanism that preserves edges and can be generalized to any dimension. In [198, 118], methods for constraining the evolution and the construction of convergent schemes based on the knowledge of the noise variance are reported.

10.3 Movies and Volumetric Medical Images

Traditionally, MRI volumetric data are referred to as 3D medical image. Following our framework, a more appropriate definition is of a 3D manifold in 4D (x, y, z, I). In a very similar manner we will consider gray-level movies as a 3D manifold in 4D, where all we need to do is the mental exercise of replacing z of the volumetric medical images by the sequence (time) axis. This is a relatively simple case, since now we have co-dimension equal to one.

The line element is

$$ds^2 = dx^2 + dy^2 + dz^2 + dI^2.$$

The induced metric in this case is given by

$$(g_{\mu\nu}) = \begin{pmatrix} 1 + I_x^2 & I_x I_y & I_x I_z \\ I_x I_y & 1 + I_y^2 & I_y I_z \\ I_x I_z & I_y I_z & 1 + I_z^2 \end{pmatrix}, \tag{10.6}$$

and the Beltrami flow is

$$I_t = \frac{1}{\sqrt{g}} \text{div} \left(\frac{\nabla I}{\sqrt{g}} \right), \tag{10.7}$$

where now $\nabla I \equiv (I_x, I_y, I_z)$ and $g = 1 + I_x^2 + I_y^2 + I_z^2$.

The meaning of edge preserving in movies is as follows: In a shot where things stay more or less in the same place, the flow tends to flatten the boundaries, that is, it is a "steady shot" filter. Yet it does not have an impact on an adjacent different scene since it preserves sharp changes along the time axis.

10.4 The Image Area as a Measure for Color Processing

We show in this section that the geometric framework results in a meaningful operator for enhancing color images. The area functional, or "norm," captures the way we would like the smoothing process to act on the different color channels while exploring the coupling between them. Next, the steepest descent flow associated with the first variation of this functional is shown to be a proper selective smoothing filter for the color case. In this section we justify the usage of the area norm and the Beltrami steepest descent flow in the color case. We list the requirements, compare to other norms, and relate to line element theories in color.

In [197, 198, 118, 108], minimizing the area of the image surface is claimed to yield a simple filter for color image enhancement. The area norm may serve for intermediate asymptotic analysis in low-level vision, which is referred to as "scale space" in the computer vision community [168]. The norm may be coupled with variance constraints that are implemented via projection methods that were used for convergence-based denoising [173] for image processing. Another popular option is to combine the norm with lower-dimensional measures to create variational segmentation procedures, like the Mumford–Shah [153]. In this section we justify the usage of the area norm for color images obtained by the geometric framework and the Beltrami flow as its scale space.

We limit our discussion to variational methods in nonlinear scale-space image processing, and to Euclidean color space. Given other significant group of transformations in color, one could design the invariant flow with respect to that group based on the philosophy of images as surfaces in the hybrid space (x, y, R, G, B) through an arclength definition.

Let us first explore the geometric framework relation to line element theory in color. Next, we list the coupling requirements for the color case. A simple "color image formation" model defines a proper order of events for a desired enhancement. It is shown that this sequence of events is captured by the area norm.

10.4.1   The Geometric Framework and Color Processing

Usually, a color image is considered as three images red, green, and blue, that are composed into one. How should we treat such a composition? To answer this question, we view color images as *embedding maps* that flow toward *minimal surfaces*. See [219] for a nonvariational related effort.

At this point we would like to go back more than a hundred years, when physicists started to describe the human color perception as simple geometric space. Helmholtz [207] was the first to define a "line element" (arclength) in color space. He first used a Euclidean R, G, B space defined by the arclength

$$ds^2 = (d \log R)^2 + (d \log G)^2 + (d \log B)^2. \tag{10.8}$$

His first model failed to represent empirical data of human color perception. Schrödinger [183] fixed the Helmholtz model by introducing the arclength

$$ds^2 = \frac{1}{l_R R + l_G G + l_B B} \left(\frac{l_R (dR)^2}{R} + \frac{l_G (dG)^2}{G} + \frac{l_B (dB)^2}{B} \right), \tag{10.9}$$

where l_R, l_G, l_B are constants. Schrödinger's model was later found to be inconsistent with findings on threshold data of color discrimination.

If we summarize the existing models for color space, we have two main cases: (1) the *inductive* line elements that derive the arclength by simple assumptions on the visual response mechanisms. For example, we can assume that the color space can be simplified and represented as a Riemannian space with zero Gaussian curvature like the Helmholtz [207] or Stiles [202, 217] models. Another possibility for inductive line elements is to consider color arclengths like Schrödinger, or Vos–Walraven [208]. These models define color spaces with nonzero curvature ("effective" arclength). (2) the *empirical* line elements, in which the metric coefficients are determined to fit empirical data. Some of these models describe a Euclidean space like the CIELAB (CIE 1976 ($L^*a^*b^*$)) [217], recently used in [179]. Others, like MacAdam [139, 140], are based on an effective arclength.

The geometric framework is not limited to zero curvature spaces and can incorporate any inductive or empirical color line element. See, for example, [199].

In case we want to perform any meaningful processing operation on a given image, we need to define a spatial relation between the points in the image plane $\mathbf{x}$. As a first step define the image plane to be Euclidean, which is a straightforward assumption for 2D images, that is,

$$ds_{\mathbf{x}}^2 = dx^2 + dy^2. \tag{10.10}$$

In order to construct a valuable geometric measure for color images we need to combine the spatial and color measures. The simplest combination of this hybrid spatial-color space is given by

$$ds^2 = ds_{\mathbf{x}}^2 + \beta^2 ds_c^2. \tag{10.11}$$

The parameter β has dimensions [distance/intensity] and fixes the relative scale between the intensity of colors and the spatial distances. For a large β it defines a regularization of the color space.

Given the above arclength for color images, we pose the following question: How should a given image be simplified? In other words: What measure/norm/functional is meaningful? What kind of variational method should be applied in this case?

Once we defined the arclength we can measure area. Area minimization is a well-known and studied physical phenomena, and indeed for the right aspect ratio β, the area is a meaningful measure for our color case. Once the minimization measure is determined, one still needs to determine the parameterization for the steepest decent flow. The geometric flow for area minimization, which preserves edges the most, is given by the Beltrami flow.

Let x and y be the *spatial* coordinates and the intensity R, G, B the *feature* coordinates, and describe color images as 2D surfaces in the 5D

(x, y, R, G, B) space. The arclength is given by

$$ds^2 = dx^2 + dy^2 + dR^2 + dG^2 + dB^2. \tag{10.12}$$

As an introduction we have chosen the oversimplified Euclidean color space, and for the time being assume $\beta = 1$. Next, we *pull back* the image surface-*induced metric* from the arclength definition. By applying the chain rule $dR = R_x dx + R_y dy$, and rearranging terms, we obtain a distance measure on the surface defined via

$$ds^2 = g_{11} dx^2 + 2g_{12} dx dy + g_{22} dy^2,$$

where $g_{\mu\nu} = \delta_{\mu\nu} + \sum_i I_\mu^i I_\nu^i$ are the induced metric coefficients, $i \in \{1, 2, 3\}$ indicates the different color channels: $I^1 = R$, $I^2 = G$ and $I^3 = B$.

For the Euclidean color case with the induced metric, the norm is the area $\int d^2 x \sqrt{g}$. Here g is the determinant of the metric matrix $g = \det(g_{ij}) = g_{11}g_{22} - g_{12}^2$ given by its components $g_{\mu\nu} = \delta_{\mu\nu} + \sum_i I_\mu^i I_\nu^i$. If we multiply the intensities by a constant β, this functional is given explicitly by

$$S = \int \sqrt{1 + \beta^2 \sum_i |\nabla I^i|^2 + \beta^4 \frac{1}{2} \sum_{ij} (\nabla I^i, \nabla I^j)^2} \, dx dy, \tag{10.13}$$

where $(\nabla R, \nabla G) \equiv R_x G_y - R_y G_x$ is the magnitude of the cross product of the vectors ∇R and ∇G. The functional in Eq. (10.13) is the area of the image as a surface.

This functional obviously depends on the scalar β. For $\beta \gg 1$ it practically means mapping the intensity values that usually range between 0 and 255 to, let's say, $[0, 1000]$. Roughly speaking, for this limit of β, the order of events along the scale of the flow is as follows: First the different colors align together, then the selective smoothing geometric flow starts (similar to the single channel TV-L_1). On the other limit, where $\beta^2 \ll 1$, the smoothing tends to occur uniformly as a multichannel heat equation (L_2).

10.4.2   Color Image Formation and Coupling Requirements

Let us elaborate on the selection of area as a proper measure for color images. The question we try to answer is how should we link between the different spectral channels. Let us assume that each color is "equally important" and thus the measure we define should be symmetric. Within the scale-space philosophy, we want the different spectral channels to get smoother in scale. This requirement leads to the minimization of the different color channels gradient magnitudes combined in one way or another.

Next, we argue that an important demand for color image processing is the alignment requirement of the different color channels. That is, we want the color channels to align together as they become smoother in scale.

Figure 10.3 shows one level set of the red and green colors and their gradient vectors at one point along their corresponding level sets. The requirement that the color channels align together as they evolve amounts to minimizing the cross products between their gradient vectors.

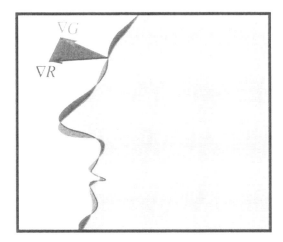

Figure 10.3: The cross product between ∇R and ∇G, $(\nabla G, \nabla R)/2$ displayed as the area of the gray triangle, measures the alignment between the channels.

A simplified color image formation model is a result of viewing Lambertian surface patches (not necessarily flat). Such a scene is a generalization of a "Mondriaan world," named after the artist Piet Mondriaan. Each channel is considered as the projection of the real 3D world surface normal $\hat{\mathbf{N}}(\mathbf{x})$ onto the light source direction $\vec{l}$, multiplied by the albedo $\rho(x, y)$. The albedo captures the characteristics of the 3D object's material and is different for each spectral channel. The three color channels may then be written as

$$I^i(\mathbf{x}) \;=\; \rho_i(\mathbf{x})\hat{\mathbf{N}}(\mathbf{x}) \cdot \vec{l}; \tag{10.14}$$

see Figure 10.4. This means that the different colors capture the change in material via the albedo that multiplies the normalized shading image $\tilde{I}(\mathbf{x}) = \hat{\mathbf{N}}(\mathbf{x}) \cdot \vec{l}$.

Let us also assume that the material, and therefore the albedo, are the same within a given object in the image (e.g., $\rho_i(\mathbf{x}) = c_i$, where c_i is a given constant). The intensity gradient for each channel within a given object is then given by

$$\begin{aligned}
\nabla I^i(\mathbf{x}) &= \tilde{I}(\mathbf{x})\nabla\rho_i(\mathbf{x}) + \rho_i(\mathbf{x})\nabla\tilde{I}(\mathbf{x}) \\
&= \tilde{I}(\mathbf{x})\nabla c_i + c_i\nabla\tilde{I}(\mathbf{x}) \\
&= c_i\nabla\tilde{I}(\mathbf{x}).
\end{aligned} \tag{10.15}$$

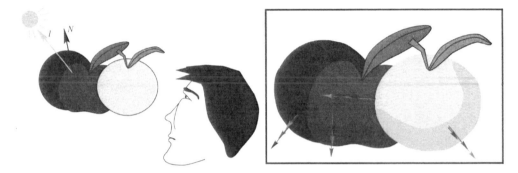

Figure 10.4: A simplified Lambertian color image formation model (left) leads to spectral channel alignment (right).

Under the above assumptions, all color channels should have the same gradient direction within a given object. Moreover, the gradient direction should be orthogonal to the boundary for each color, since both the normalized shading image $\tilde{I}$ and the albedo ρ_i change across the boundaries. We can thus conclude that a first step in color processing should be the alignment of the colors so that their gradients agree. Only next should come the diffusion of all the colors simultaneously.

For a large enough β, Eq. (10.13) follows exactly these requirements and the area norm is a regularization form of

$$\int \sqrt{\sum_i |\nabla I^i|^2 + \beta^2 \sum_{ij} (\nabla I^i, \nabla I^j)^2} dx dy, \qquad (10.16)$$

which captures the order of events described above. For an even larger β, it can be considered as a regularization of the *affine invariant* norm

$$\int \sqrt{\sum_{ij} (\nabla I^i, \nabla I^j)^2} dx dy. \qquad (10.17)$$

If we also add the demand that edges should be preserved and search for the simplest geometric parameterization for the flow, we end up with the Beltrami flow as a proper selection.

Figure 10.5 shows snapshots from the Beltrami scale-space in color for three images. Next, the flow is used to selectively smooth the JPEG compression distortions in Figure 10.6. Observe how the color perturbations are smoothed: The cross correlation between the colors holds the edges while selectively smoothing the noncorrelated data.

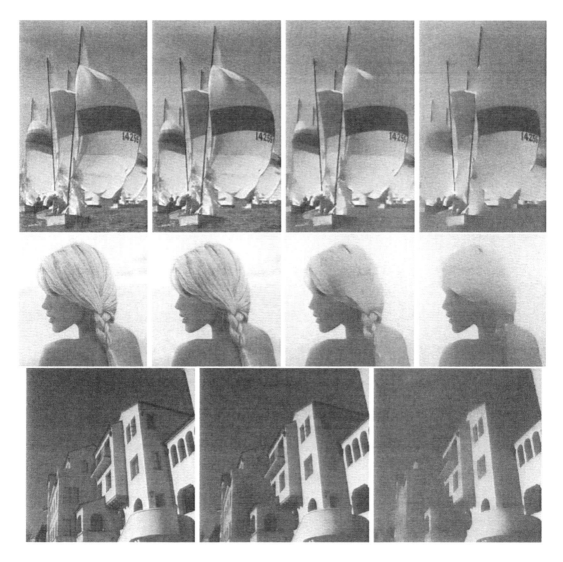

Figure 10.5: Snapshots along the scale space (leftmost is the original image).

Other norms for multispectral image processing and scale-space filters are proposed in [38, 179, 209, 211, 175, 15, 194]. We have shown that the geometric framework yields a proper norm with respect to recent norms, and with respect to a list of objective requirements and considerations of color image formation. Next, we apply the Beltrami operator to construct an orientation-preserving flow for texture images.

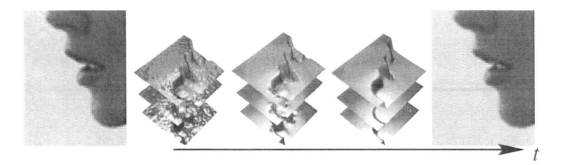

Figure 10.6: Three snapshots along the scale space for selectively smoothing JPEG lossy effects. The three channels are rendered as surfaces. The original image is on the left.

10.5   The Metric as a Structure Tensor

In [73, 134], Gabor considers an image enhancement procedure based on a single numerical step along a directional flow. It is based on the anisotropic flow via the *inverse* second directional derivative in the "edge" direction (∇I direction) and the geometric heat equation (second derivative in the direction parallel to the edge). The same idea of steering the diffusion direction motivated many recent works.[2] Cottet and Germain [50] use a smoothed version of the image to direct the diffusion, while Weickert [212, 210] also smooths the "structure tensor" $\nabla I \nabla I^T$ and then manipulates its eigenvalues to steer the smoothing direction. Eliminating one eigenvalue from a structure tensor, first proposed as a color tensor in [56], is used in in [179, 178], in which the tensors are not necessarily positive definite, while in [209, 211], the eigenvalues are manipulated to result in a positive definite tensor. See also [38], where the diffusion is in the direction perpendicular to the maximal gradient of the three color channels.

Motivated by all of these results we will first link the anisotropic orientation diffusion (coherence enhancement) to the geometric framework, and then invert the diffusion direction across the edge. Let us first show that the diffusion directions can be deduced from the smoothed metric coefficients $g_{\mu\nu}$ and may thus be included within the Beltrami framework under the right choice of directional diffusion coefficients.

The induced metric ($g_{\mu\nu}$) is a symmetric uniformly positive definite matrix that captures the geometry of the image surface. Let λ_1 and λ_2 be

---

[2]This definition of anisotropic flow differs from the Perona–Malik [164] framework, which is locally isotropic. See [169] for many interesting extensions and applications of the locally isotropic flow.

the largest and the smallest eigenvalues of $(g_{\mu\nu})$, respectively. Since $(g_{\mu\nu})$ is a symmetric positive matrix, its corresponding eigenvectors u_1 and u_2 can be chosen orthonormally. Let $\mathbf{U} \equiv (u_1|u_2)$, and $\Lambda \equiv \begin{pmatrix} \lambda_1 & 0 \\ 0 & \lambda_2 \end{pmatrix}$; then we readily have the equality

$$(g_{\mu\nu}) = \mathbf{U}\Lambda\mathbf{U}^T. \tag{10.18}$$

Note also that

$$(g^{\mu\nu}) \equiv (g_{\mu\nu})^{-1} = \mathbf{U}\Lambda^{-1}\mathbf{U}^T = U \begin{pmatrix} 1/\lambda_1 & 0 \\ 0 & 1/\lambda_2 \end{pmatrix} \mathbf{U}^T, \tag{10.19}$$

and that

$$g \equiv \det(g_{\mu\nu}) = \lambda_1\lambda_2. \tag{10.20}$$

We will use the image metric in its geometric interpretation, that is, as a structure tensor. The coherence enhancement Beltrami flow $\mathbf{I}_t = \Delta_{\hat{g}}\mathbf{I}$ for color-texture images is then given as follows:

1. Compute the metric coefficients $g_{\mu\nu}$. For the N channel case (for color $N = 3$) we have

$$g_{\mu\nu} = \delta_{\mu\nu} + \sum_{k=1}^{N} I_\mu^k I_\nu^k. \tag{10.21}$$

2. Diffuse the $g_{\mu\nu}$ coefficients by convolving with a Gaussian of variance ρ, thereby

$$\tilde{g}_{\mu\nu} = G_\rho * g_{\mu\nu}. \tag{10.22}$$

For 2D images $G_\rho = (1/\pi\rho^2)e^{-(x^2+y^2)/\rho^2}$.

3. Change the eigenvalues, λ_1, λ_2, $\lambda_1 > \lambda_2$, of $(\tilde{g}_{\mu\nu})$ so that $\lambda_1 = \alpha^{-1}$ and $\lambda_2 = \alpha$, for some given positive scalar $\alpha \ll 1$. This yields a new metric $\hat{g}_{\mu\nu}$ that is given by

$$(\hat{g}_{\mu\nu}) = \tilde{\mathbf{U}} \begin{pmatrix} \alpha^{-1} & 0 \\ 0 & \alpha \end{pmatrix} \tilde{\mathbf{U}}^T = \tilde{\mathbf{U}}\Lambda_\alpha\tilde{\mathbf{U}}^T. \tag{10.23}$$

4. Evolve the kth channel via Beltrami flow, that by the selection $\hat{g} \equiv \det(\hat{g}_{\mu\nu}) = \lambda_1\lambda_2 = \alpha^{-1}\alpha = 1$ now reads

$$\begin{aligned} I_t^k &= \Delta_{\hat{g}}I^k \equiv \frac{1}{\sqrt{\hat{g}}}\partial_\mu\sqrt{\hat{g}}\hat{g}^{\mu\nu}\partial_\nu I^k = \partial_\mu\hat{g}^{\mu\nu}\partial_\nu I^k \\ &= \operatorname{div}\left(\tilde{\mathbf{U}} \begin{pmatrix} \alpha & 0 \\ 0 & \alpha^{-1} \end{pmatrix} \tilde{\mathbf{U}}^T\nabla I^k\right) \\ &= \operatorname{div}\left(\tilde{\mathbf{U}}\Lambda_\alpha\tilde{\mathbf{U}}^T\nabla I^k\right). \end{aligned} \tag{10.24}$$

Note again that both for gray-level and color images the above flow is similar to the coherence-enhancing anisotropic diffusion with the important property of a uniformly positive definite diffusion tensor. For color images, $(g_{\mu\nu}) = \mathcal{I} + \sum_i \nabla I^i \nabla I^{iT}$, where $\mathcal{I}$ is the identity matrix, and I^i are the color channels $((I^r, I^g, I^b) \equiv (I^1, I^2, I^3))$. In this case all that is done is the identity added to the structure tensors $\nabla I \nabla I^T$ for gray and $\sum_i \nabla I^i \nabla I^{iT}$ for color. This addition does not change the eigenvectors and thus the above flow is equivalent to Weickert schemes [212, 210, 209, 211]. Next, we introduce an inverse/direct diffusion model that takes us beyond the selective smoothing into sharpening.

10.6 Inverse Diffusion Across the Edge

Let us take one step further, and exit our Riemannian framework by defining $(g_{\mu\nu})$ to be a nonsingular symmetric matrix with one positive and one *negative* eigenvalues, that is, a pseudo-Riemannian metric. Instead of a small diffusion we introduce an inverse diffusion across the edge. Here we extend Gabor's idea [73, 134] of inverting the diffusion along the gradient direction.

Inverting the heat equation is an inherently unstable process; see, for example, [201]. However, if we keep smoothing the metric coefficients, and apply the heat operator in the perpendicular direction, we get a coherence-enhancing flow with sharper edges that is stable for a short duration of time.

The idea is simply to change the sign of one of the modified eigenvalues in the algorithm described in the previous section; see Figure 10.7.

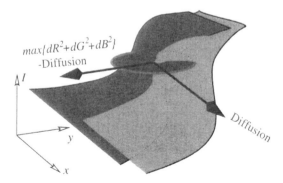

Figure 10.7: Motivated by the geometric framework and Gabor's sharpening algorithm we steer the diffusion directions and invert the diffusion direction across the edge. The edge direction is extracted by "sensing" the multichannel structure after smoothing the metric.

We change steps 3 and 4 of the previous scheme, which now reads

1. Compute the metric coefficients $g_{\mu\nu} = \delta_{\mu\nu} + \sum_{k=1}^{N} I_{\mu}^{k} I_{\nu}^{k}$.

2. Diffuse the $g_{\mu\nu}$ coefficients by convolving with a Gaussian of variance ρ.

3. Change the eigenvalues of $(\tilde{g}_{\mu\nu})$ such that the largest eigenvalue λ_1 is now $\lambda_1 = -\alpha^{-1}$ and $\lambda_2 = \alpha$, for some given positive scalar $\alpha < 1$. This yields a new *matrix* $\hat{g}_{\mu\nu}$ that is given by

$$(\hat{g}_{\mu\nu}) = \tilde{\mathbf{U}} \begin{pmatrix} -\alpha^{-1} & 0 \\ 0 & \alpha \end{pmatrix} \tilde{\mathbf{U}}^T = \tilde{\mathbf{U}} \Lambda_\alpha \tilde{\mathbf{U}}^T. \tag{10.25}$$

We have used a single scalar α for simplicity of the presentation. Different eigenvalues can be chosen; one example is eigenvalues that depend on the original ones and bring us closer to the Beltrami flow. By manipulating the eigenvalues we control the direction as well the intensity of the diffusion that can just as well be edge dependent. In this application the key idea is to modify the largest eigenvalue to be negative. This modification inverts the diffusion direction across the multispectral edge and thereby enhances it.

4. Evolve the kth channel via the flow, that by the selection $|\hat{g}| \equiv |\det(\hat{g}_{\mu\nu})| = |\lambda_1 \lambda_2| = |-\alpha^{-1}\alpha| = 1$, reads

$$\begin{aligned} I_t^k &= \frac{1}{\sqrt{|\hat{g}|}} \partial_\mu \sqrt{|\hat{g}|} \hat{g}^{\mu\nu} \partial_\nu I^k &= \partial_\mu \hat{g}^{\mu\nu} \partial_\nu I^k \\ &&= \operatorname{div}\left(\tilde{\mathbf{U}} \begin{pmatrix} -\alpha & 0 \\ 0 & \alpha^{-1} \end{pmatrix} \tilde{\mathbf{U}}^T \nabla I^k \right). \end{aligned}$$

For the gray-level case with $\rho = 0$ it simplifies to a highly unstable inverse heat equation. However, as ρ increases, the smoothing along the edges becomes fundamental and the scheme is similar in its spirit to that of Gabor's [73]. He comments on the inverse diffusion operation in the gradient direction is that "*It is very similar to the operation which the human eye carries out automatically, and it is not surprising that even the first steps in imitating the human eye by mechanical means lead to rather complicated operations.*" It is important to note that the idea of stabilizing the inverse heat equation is extensively used in image processing. Exploring this area is beyond the scope of this chapter. However, we would like to refer the reader to the "shock filters" introduced by Osher and Rudin in [160] for gray-level images, and the extension of Alvarez and Mazorra [5] who apply geometrical inverse diffusion in the gradient direction combined with a directional smoothing in the orthogonal direction for gray-level images.

10.6.1   Color Orientation-Sharpening Examples

In [212] the coherence enhancement flow is applied on several masterpieces by van Gogh, which results in a "coherence enhancement of expressionism." In the next example we have chosen to "enhance and sharpen impressionism." We first apply the anisotropic oriented diffusion flow and then the new oriented diffusion along/inverse diffusion across the edge on a color painting by Claude Monet; see Figure 10.8.

Figure 10.8: Original picture "Femme à l'ombrelle tournée vers la gauche," by Claude Monet 1875 ("woman with umbrella turning left") 521 × 784 (left), and the result of the inverse/direct diffusion flow ($\rho = 4$) for 8 iterations (right).

Next, we apply the color-oriented diffusion and the oriented inverse/direct diffusion algorithms to standard color-texture test images in Figures 10.9 and 10.10.

Figure 10.9: Color and texture: original "Shells" image 242×184 (left), and the result of the color and texture inverse/direct diffusion.

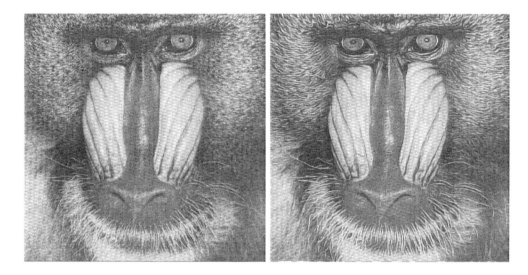

Figure 10.10: Color and texture: Diffusion flow with smoothed metric and steered eigenvalues ($\alpha = 10^{-5}$); original "mandrill" image 512×512 (left), and the result of orientation-preserving flow and negative eigenvalue (inverse diffusion) in gradient direction, $\alpha = 0.39$.

10.7 Summary

The geometric framework can be used to design novel procedures for enhancement of color and texture images. These procedures are based on the interpretation of the image as a surface and a geometric heat flow with respect to a given metric (Beltrami operator) as a filter. We linked the geometric framework to color and texture enhancement algorithms and explored a sharpening procedure, based on inverse diffusion across the edge, that was first introduced in [108, 107]. Armed with this tool, a direct application of the Beltrami flow is to enhance, selectively smooth, or sharpen color-texture and volumetric images. It can be used to reduce the image entropy with a minor perceptual reduction prior to compression. It is useful for image coherence enhancement as part of decoding and reconstruction process, for example, image restoration and denoising of lossy compression effects.

10.8   Exercises

1. Prove that the area functional $\int da = \int \sqrt{g}dxdy$ for a color image as a surface in a Euclidean space $(x, y, \beta R, \beta G, \beta B)$ can be written as Eq. (10.13).

2. Prove that $\Delta_g \mathbf{I} = 0$ is indeed the Euler–Lagrange equation for the area $\int dx^2 \sqrt{g}$.

3. Prove that one of the eigenvectors of (g_{ij}) is along the $\max\{dR^2 + dG^2 + dB^2\}$ direction.

4. Apply the inverse diffusion to color images. Test different choices for α.

5. Research project on color video: Show that the Beltrami flow increases the "compressibility" of a movie.

6. Research project: What is the right scale? Is there a direct analytical relation among σ, the variance of the noise, and the time t we have to evolve the TV/Beltrami flows? Explore the link between the probability to find a random walk particle at a given location, the Gaussian distribution, and the heat equation. Next extend to random walks on curved manifolds, and eventually to random walks on dynamic manifolds.

Texture Mapping, Matching Isometric Surfaces, and 3D Face Recognition

In the preface of this book we pointed out our natural interest in image analysis, a task that engages most of the human brain's cortical activity. One fascinating property of the human mind is its sensitivity and ability to recognize human faces. That is, we are trained, quite successfully too, to recognize (a finite set of) faces. We can recognize a known face under a wide range of lighting conditions. Moreover, the variety of facial expressions displayed by humans does not influence our recognition performances to any large extent. An interesting fact is that humans (unlike monkeys) are quite sensitive to the facial orientation. The question of how we do this preoccupies computer vision researchers and is a major challenge many have now set out to address. It is believed that once the way this task is done by the brain is well understood, automatic face recognition methods will follow naturally. Although many solutions have been proposed [110], so far all commercial systems fail to reliably recognize people based on images of their faces.

Today (the end of 2003), all commercial face recognition systems aim to mimic the human recognition system. Recognition is therefore based on images, or a sequence of images of the subject. In this chapter, we discuss a somewhat different approach that we have recently tested [22]. Our challenge was to distinguish between two identical twins under varying facial expressions. For this delicate recognition task we first capture the three-dimensional structure of the face. Then, a signature, invariant to facial expressions, is computed, and the recognition is performed by signature matching; see Figure 11.1.

It is interesting to note that the human visual system is clearly not sensitive to the geometry of facial surface, and focuses on facial texture

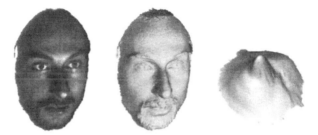

Figure 11.1: The facial surface is captured with a 3D scanner. Then a bending-invariant signature is computed.

rather than on geometric structure. Taking advantage of this, people wearing makeup can fool others' perceptions of their geometric facial structure.

Before tackling the three-dimensional face recognition problem directly, we shall deal with two related problems: texture mapping in computer graphics, and isometric surface matching in shape analysis. We use the fast marching method on triangulated domains, presented in Chapter 7 as one step that helps us flatten the surface. We then use the output for texture mapping, surface matching, and finally for 3D face recognition.

This chapter describes work published in a series of recent papers, starting with a joint work with Gil Zigelman (my first graduate student) and Nahum Kiryati [221], in which the problem of texture mapping was solved by flattening the surface into a plane. Next, together with Asi Elad [63, 64], we dealt with matching isometric surfaces. There, surfaces are flattened into higher dimensional Euclidean spaces such as $\mathbb{R}^3$ or $\mathbb{R}^4$. Finally, together with Alex and Michael Bronstein [22], we showed how ideas from [63, 64] can be modified and integrated into a face recognition system that can successfully distinguish between identical twins.

11.1 Flat Embedding

Flat embedding is the process of mapping a given set of surface points, with given relative surface (geodesic) distances between each two points in this set, into a finite-dimensional Euclidean space. Flat embedding of a smooth surface into a plane, also known as the "map-maker problem", was shown to be a useful operation in the analysis of cortical surfaces [185], and texture mapping [185, 221, 84]. While embedding into a plane provides a parameterization of the surface that can be used for texture mapping, flat embedding into higher dimensional Euclidean spaces provides an efficient way to compute bending-invariant signatures for isometric surfaces [63]. It transforms the matching operation of nonrigid objects into a much simpler problem of matching rigid ones, and was found to be especially effective for 3D face recognition [22].

We refer to "flattening" or "flat embedding" of a surface as the process of mapping a curved surface $\mathcal{S}_1$, into a surface $\mathcal{S}_2$, embedded in $\mathbb{R}^m$. The flattening process maps every pair of surface points p_i, p_j in $\mathcal{S}_1$, with a minimal geodesic distance $\delta_{ij} = d_g(p_i, p_j)$ between them, into the $\mathcal{S}_2$ surface points q_i, q_j, with the Euclidean distance $d_{ij} = \|q_i - q_j\|$ between the points. Here we assume that q_i and q_j also indicate the coordinates of the points in the Euclidean embedding space. The mapping is said to be "flat" if the geodesic distance δ_{ij} is as close as possible to the corresponding Euclidean distance d_{ij} of the corresponding points in the dual surface $\mathcal{S}_2$. The mapping distortion is measured in a global integral manner (L_2), accumulating the error for every corresponding pair of points, unlike Lipschitz embedding that measures the largest relative error (L_∞).

The steps we follow in all applications are as follows. First, geodesic distances are computed between points uniformly scattered on a given triangulated surface. The resulting matrix of relative geodesic distances, $\mathbf{M}$, is then given by its elements $M_{ij} = \delta_{ij}^2$. Next, the coordinates of the points in $\mathbb{R}^m$ are computed by minimizing some integral measure of the local errors, such as $\sum_{ij}(\delta_{ij} - d_{ij})^2$. The final step is determined by the application. When we deal with texture mapping, the embedding space is $\mathbb{R}^2$, in which case the flattened surface coincides with the embedding space. Whereas for isometric surface matching, the embedding space is usually $\mathbb{R}^3$ or a higher dimensional space. Before we demonstrate the applications, let us present the dimensionality reduction procedure we use, namely, classical scaling.

11.1.1 Classical Scaling

Multi-Dimensional Scaling or MDS is a family of methods that tries to embed a set of "feature items" as points in a finite-dimensional small Euclidean space. For some books on this topic, see [123, 53, 18]. The embedding is such that the distance between every two items of the given set is approximated as best as possible by the Euclidean distance between the two corresponding points in the embedding Euclidean space. The approximation measure, and the numerical method used for finding the coordinates of the points in the Euclidean space, define the specific MDS method. In this chapter we deal only with "Classical Scaling" due to its simplicity and its nice property of relating the algebraic structure of the distance matrix to geometry.

Classical scaling was used in the context of image analysis by Rubner and Tomasi [172], who applied it to texture classification. They define metric perceptual similarities between textures, and use MDS in order to visualize the different texture items ordered in a plane or in 3D space. Such applications fall into the standard usage of MDS methods, that is, using some arbitrary measurements of distance between feature items to lower the dimensionality in which this distance is measured, so that the data can be better viewed, clustered, or analyzed.

MDS methods were first used to flatten surfaces into a plane by Schwartz et al. in [185]. Their work was, in some sense, a breakthrough, in which surface geometry was translated into a plane. However, flat embedding into a plane was not enough to match convoluted surfaces. The plane restriction introduces deformations that actually prevented proper matching of convoluted surfaces. This problem can be solved if higher dimensions of the embedding space are considered. This region was discovered, explored, and utilized to solve the surface matching problem in [63, 64].

Classical scaling was invented by Young et al. in the 1930s. They showed that given a matrix of distances between points in a Euclidean space, it is possible to extract their coordinates such that the distances are preserved. When adapted to our applications, it allows us to evaluate, in a simple way, the representation error for a given surface after flat embedding in a given number of dimensions.

Let the coordinates of n points in a k-dimensional Euclidean space $\mathbb{R}^k$ be given by $\mathbf{p}_i$, $(i = 1, ..., n)$, where $\mathbf{p}_i = [p_{i1}, p_{i2}, ..., p_{ik}]^T$. The matrix $\mathbf{P}_{n \times k}$ represents the coordinates of these n points in $\mathbb{R}^k$. The Euclidean distance between points i and j is given by

$$
\begin{aligned}
d_{ij}^2 = \sum_{l=1}^{k} (p_{il} - p_{jl})^2 &= [\mathbf{p}_i - \mathbf{p}_j]^T [\mathbf{p}_i - \mathbf{p}_j] \\
&= \mathbf{p}_i^T \mathbf{p}_i - 2\mathbf{p}_i^T \mathbf{p}_j + \mathbf{p}_j^T \mathbf{p}_j \\
&= |\mathbf{p}_i|^2 - 2\mathbf{p}_i^T \mathbf{p}_j + |\mathbf{p}_j|^2.
\end{aligned}
\tag{11.1}
$$

where T denotes the transpose operator.

Define the matrix $\mathbf{E}$ elements to be the square Euclidean distances between each pair of points, that is, $E_{ij} = d_{ij}^2$. Then, $\mathbf{E}$ can be compactly written as

$$
\mathbf{E} = \mathbf{Q} - 2\mathbf{P}\mathbf{P}^T + \mathbf{Q}^T,
\tag{11.2}
$$

where $\mathbf{Q} = \mathbf{c}\mathbf{1}^T$, and $\mathbf{1}$ is a ones vector of length n, and $\mathbf{c}$ is a vector of length n in which $c_i = |\mathbf{p}_i|^2 = \sum_{l=1}^{k} p_{il}^2$.

Consider the translation of the origin of the points defined by $\mathbf{P}$ to a new location. Define this location to be an affine combination of the points themselves: $\mathbf{s}^T = \mathbf{w}^T \mathbf{P}$, where $\sum_{i=1}^{n} w_i = 1$, and for every i, $w_i \geq 0$. Then, the coordinates with respect to the new origin, denoted by the matrix $\tilde{\mathbf{P}}$, are given by

$$
\begin{aligned}
\tilde{\mathbf{P}} &= \mathbf{P} - \mathbf{1}\mathbf{s}^T \\
&= (\mathbf{I} - \mathbf{1}\mathbf{w}^T)\mathbf{P} \\
&= \mathbf{O}_w \mathbf{P}.
\end{aligned}
\tag{11.3}
$$

Multiplying $\mathbf{O}_w$ by a vector of ones on either side yields a vector of zeros,

$$\mathbf{O}_w \mathbf{1} = (\mathbf{1}^T \mathbf{O}_w)^T = \mathbf{0}.$$

If we choose $\mathbf{s}$ to be the center of mass of the points, by using

$$\mathbf{w} \;=\; \frac{1}{n}\mathbf{1},$$

the corresponding $\mathbf{O}_w$ is denoted by $\mathbf{J}$. It is defined by

$$\mathbf{J} \;=\; \mathbf{I} - \frac{1}{n}\mathbf{11}^T, \tag{11.4}$$

where $\mathbf{I}$ is the identity matrix. $\mathbf{J}$ is called a centering matrix. If $\mathbf{P}$ is already column centered, that is, the center of mass of the points defined by $\mathbf{P}$ is the origin of their coordinates, then

$$\mathbf{JP} \;=\; \mathbf{P}. \tag{11.5}$$

Define the inner product matrix to be $\mathbf{B}$, for which the ij element is given by $B_{ij} = \mathbf{p}_i^T \mathbf{p}_j$. That is, $\mathbf{B} = \mathbf{PP}^T$, where again $\mathbf{P} = [\mathbf{p}_1, \ldots, \mathbf{p}_n]^T$ is the $n \times k$ matrix of the coordinates. Then, the squared distances matrix $\mathbf{E}$ can be written as

$$\mathbf{E} \;=\; \mathbf{Q} - 2\mathbf{B} + \mathbf{Q}^T.$$

Thereby, given the squared distances matrix $\mathbf{E}$, the inner product matrix can be extracted by

$$\mathbf{B} \;=\; -\frac{1}{2}\mathbf{JEJ}. \tag{11.6}$$

By double centering, the centering matrix $\mathbf{J}$ eliminates the $\mathbf{Q}$ matrices by subtracting the average of each row or column from its elements. Next, if the coordinates of the points are extracted from $\mathbf{B}$, the centering matrix $\mathbf{J}$ sets the origin to be the center of mass of the points.

The inner product matrix $\mathbf{B}$ is symmetric, positive semi-definite, and of rank k. Therefore, $\mathbf{B}$ has k nonnegative eigenvalues and $n - k$ zero eigenvalues. The matrix $\mathbf{B}$ can be written in terms of its spectral decomposition as

$$\mathbf{B} \;=\; \mathbf{V\Lambda V}^T, \tag{11.7}$$

where

$$\Lambda_{n \times n} \;=\; \mathrm{diag}(\lambda_1, \lambda_2, \ldots, \lambda_k, 0, \ldots, 0).$$

For convenience, the eigenvalues of $\mathbf{B}$ are ordered such that $\lambda_1 \geq \lambda_2 \geq \cdots \geq \lambda_k \geq 0$. Hence, the required coordinates are given (up to rotation, translation, and reflection) by using the nonzeros sub-matrix $\Lambda_{k \times k}$ and the corresponding eigenvectors sub-matrix $\mathbf{V}_{n \times k}$,

$$\mathbf{P}_{n \times k} = \mathbf{V}_{n \times k} \Lambda_{k \times k}^{\frac{1}{2}}.$$

The classical MDS approach minimizes a version of the L_2 for matrices, also known as the Frobenius norm, given by

$$\mathcal{E}rr = \|\mathbf{V}(\Lambda - \Lambda_m)\mathbf{V}^T\|, \tag{11.8}$$

where

$$\begin{aligned} \Lambda_m &= \operatorname{diag}(\lambda_1, \lambda_2, \dots, \lambda_m, 0, \dots, 0). \\ \Lambda &= \operatorname{diag}(\lambda_1, \lambda_2, \dots, \lambda_m, \dots, \lambda_k, 0, \dots, 0), \end{aligned}$$

and $m \leq k$.

We started our discussion with distances between points in $\mathbb{R}^k$; however, the input for our problem are distances between points on a curved surface. What happens if our points are not in a Euclidean space, but on a smooth surface? This is an excellent question, for which theory still lags behind practice. Since Gauss has already shown that perfect flat embedding of curved surfaces is impossible, we do the best we can, and keep track of the embedding error. This is exactly the process of dimensionality reduction by principle component analysis (PCA). Our technique differs from PCA primarily in the geometric input that gives a completely new flavor to the problem. That is, instead of looking for the features manifold with its curved geometry, wherein today rests main effort of statistical learning, we force our distance measurement to live in a Euclidean space.

The input of our classical scaling procedure is the matrix $\mathbf{M}$ of relative geodesic distances between points. We treat $\mathbf{M}$ as if it were a realization of distances taken from a high dimensional Euclidean space, and extract, by double centering and spectral decomposition, the largest eigenvalues and corresponding eigenvectors that give us the samples of the flattened surface. Since it is impossible to flatten surfaces with effective Gaussian curvature, some of the (hopefully smaller) eigenvalues will have negative values. We sum these and get a clear estimation of the error.

11.2   Texture Mapping

The first application we deal with is texture mapping of flat texture images on arbitrary surfaces with minimal distortions. The texture mapping problem is actually the inverse problem of flattening a curved surface into a plane. We would like this mapping to preserve the local and global structures of the texture. For that goal, we design a surface flattening approach based on classical scaling that first maps the surface into a plane. Next, we map the texture image onto the curved surface while preserving the structure of the texture.

One important example of surface flattening is that of mapping the surface of the earth onto a plane, which is known as the "map-maker problem". Gauss showed in 1828 that there is no isometric mapping between two surfaces with different Gaussian curvature [58]. This means, for example, that it is impossible to map a sphere onto a plane without introducing metric distortions. Therefore, only approximate solutions for the "map-maker problem" are possible. The same is true for texture mapping in general. The main question is where and how should the deformations be introduced.

It is beyond the scope of this chapter to review the various methods for texture mapping. The problem is given here as a first application of classical scaling that operates on the result of the fast marching on triangulated domains. We point out some other interesting texture mapping methods [70, 71].

In order to use classical scaling for texture mapping we select only the two largest eigenvalues and their corresponding eigenvectors. Using these two couples we approximate the matrix $\mathbf{B} = \mathbf{JMJ} = \mathbf{V\Lambda V^T}$ as well as possible (in a least squares sense) by a matrix of rank two. The approximation error is determined by the rest of the eigenvalues, which we ignore. Classical scaling approximates the matrix $\mathbf{B}$ by a matrix of lower rank. As we saw in the previous section, $\tilde{\mathbf{P}}$ can be computed from this lower rank approximation of $\mathbf{B}$.

$\tilde{\mathbf{P}}$ minimizes the distortion function defined as

$$\mathcal{E}rr = \|\mathbf{V}(\Lambda - \Lambda_2)\mathbf{V}^T\| = \left\|\mathbf{B} - \tilde{\mathbf{P}}\tilde{\mathbf{P}}^T\right\|. \tag{11.9}$$

Low distortion mapping of a curved surface onto the plane can be used for mapping back the planar texture from the plane to the surface. After applying classical scaling to the measured geodesic distances between the surface points δ_{ij}, and extracting the two largest eigenvalues and corresponding eigenvectors via classical scaling on $\mathbf{M}$ where $M_{ij} = \delta_{ij}^2$, we get a two-dimensional flattened version of the surface. We thereby have the mapping of every point on the surface to its corresponding coordinate in the plane. Our goal now is to map the texture plane back to the surface. This procedure is straightforward. For each surface point $\mathbf{p}_i$, find the corresponding coordinate (after flattening) on the plane, $\mathbf{q}_i$. This coordinate serves as a coordinate of the texture image plane $I(\mathbf{q}_i)$. Finally, we use the point's color as the surface color at $\mathbf{p}_i$. Figures 11.2, 11.3, and 11.4 show some examples of texture mapping using the above method.

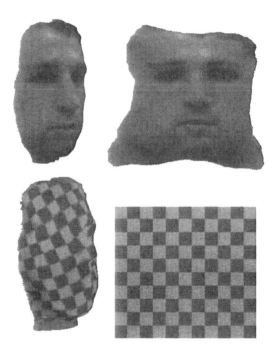

Figure 11.2: Texture mapping by flattening via MDS. Top left: Original surface. Top right: Flattened surface. Bottom right: Texture plane. Bottom left: Mapping the texture on the original surface.

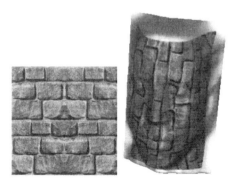

Figure 11.3: Texture mapped onto a head object by flattening the geodesic metric matrix. The global and local structures of the texture are preserved.

Figure 11.4: Chess-board texture mapping. Left: The original object. Right: The object textured by flattening the geodesic metric matrix.

11.3   Isometric Signatures for Surfaces

"An allowable mapping of the surface $\mathcal{S}$ onto the surface $\tilde{\mathcal{S}}$ is said to be *isometric* or *length preserving* if the length of any arc on $\tilde{\mathcal{S}}$ is the same as that of its inverse image on $\mathcal{S}$" [122]. Exploiting this formal definition, we would like to construct a method for matching two isometric surfaces. It is thus obvious that our method should utilize the intrinsic (geodesic) distances between surface points.

Let us start with a simple example in which we fix the flat embedding space as three-dimensional, that is $m = 3$. Given the dissimilarities matrix $\mathbf{M} = (\delta_{ij}^2)$, an MDS procedure produces coordinates, $\tilde{\mathbf{P}}$, in $\mathbb{R}^3$, for which the Euclidean distances between the points in the embedding space are as close as possible to the geodesic distances between the corresponding surface points. Given the connectivity of the vertices as triangles that represent the curved surface, we can connect the corresponding points after the MDS-mapping and obtain a surface that we refer to as a *bending-invariant signature* or *bending-invariant canonical form* of the original surface [63, 64].

The selection of $\mathbb{R}^3$ in this example was arbitrary. In order to select a "proper" dimension, m, we define the effective dimensionality to be the smallest number of dimensions that allow the invariant signature to capture a given percentage of the representation energy while being able to distinguish between different classes of surfaces in our database. Usually, it is enough for the representation error of the isometric signature surface embedded in $\mathbb{R}^m$ to be smaller than a predefined threshold. It is important, however, to recall that perfectly flat embedding is impossible for curved surfaces. This geometric deformation appears as negative eigenvalues of the double centered metric matrix. Moreover, identical positive eigenvalues introduce ambiguity of the isometric signature in the sub-space spanned by their corresponding eigenvectors. These cases rarely occur for interesting surfaces, and for any practical application, a flat embedding, for which the representation error can be numerically evaluated to be "small enough", is good enough.

We define all possible bending of a surface, which include all length preserving deformations without tearing or stretching the surface, as isometric. Let us summarize the procedure for computing bending-invariant signatures. We constructed a compact bending-invariant signature for isometric surfaces by dimensionality reduction of the metric matrix after double centering. The metric matrix entries are the squared geodesic distances between uniformly distributed points on the surface. A classical scaling procedure is applied to extract coordinates in a finite-dimensional Euclidean space in which geodesic distances are replaced by Euclidean ones. The problem of matching nonrigid objects in various postures is thereby translated into a simpler problem of matching rigid objects. In other words, we essentially transform the problem of matching isometric/nonrigid/articulated surfaces into the problem of matching rigid objects that we refer to as isometric signatures.

Figure 11.5 is a sketch of signature surfaces for two different postures of a hand. One minimal geodesic path connecting two surface points is plotted as a thick white curve, while equal geodesic distance contours from one point are plotted as thinner white contours. The minimal geodesic distance contour connecting the two points on the original hands is transformed into a straight line connecting the two corresponding points on the signature surface. We see that both hands are mapped into similar looking surfaces. These bending-invariant surfaces can be used as signatures.

There are many methods that compare rigid objects—a relatively simple problem. Some techniques try to match rigid objects by comparing their geometric-statistical properties. For example, they may compare discrete histograms of geometric measures. One simple possible matching measure for rigid objects calculates the Euclidean distances between moments of the points describing the surface. We apply this method to the signature surfaces in order to demonstrate that the main clustering effort is done by generating the bending-invariant signatures. From this point any simple classifier will finish the job.

Let M_i be a vector of the first few moments of the signature surface S_i. Then, a moments-distance matrix can be defined as

$$[\delta_M]_{ij} = \|M_i - M_j\|_2^2.$$

Applying, again, classical scaling on this matrix of distances between the moments of the flattened surfaces yields points in a new Euclidean space, where each point represents one signature surface. In this new Euclidean space, signatures of isometric surfaces are clustered together, while signatures of nonisometric surfaces are well separated.

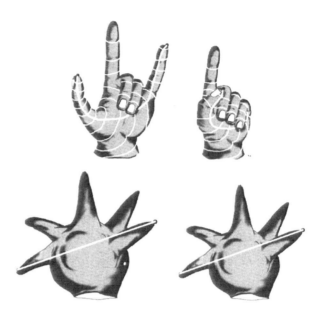

Figure 11.5: Two configurations of a hand and their corresponding isometric signatures: Top: A minimal geodesic path connecting two surface points is plotted as a thick white curve, while equal geodesic distance contours are plotted as thinner white contours. Bottom: The minimal geodesic curve on the hand becomes a straight line connecting the two corresponding points on the bending-invariant signature surface.

Figures 11.6 and 11.7 present the objects before and after the flattening procedure. Figure 11.8 shows flattening of the moments vectors into a three-dimensional Euclidean space. We see that the moments separate the objects into their classes much better after flattening.

11.4 Face Recognition

One of the most challenging problems in the field of image analysis and computer vision is face recognition. Here, following [22], we present an example of using a local version of the bending-invariant signature for 3D face recognition.

Two identical twins, Michael and Alex Bronstein, were undergraduate students attending my class in the winter of 2002. As a final project I challenged them to use the isometric signatures as a tool to distinguish between themselves under different postures and facial expressions. After

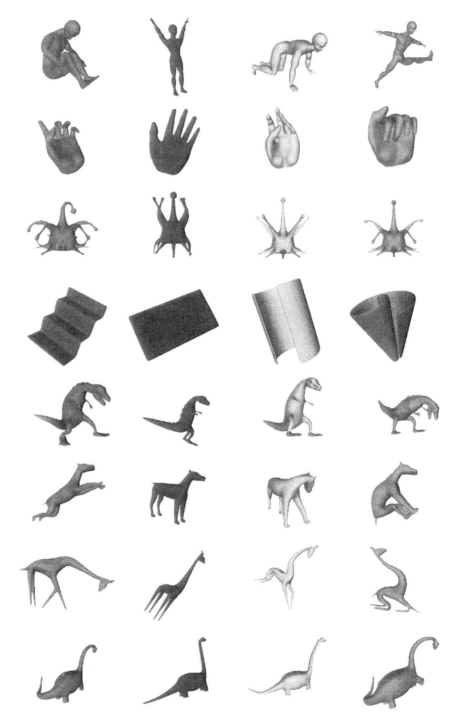

Figure 11.6: Four bending versions of eight input surfaces.

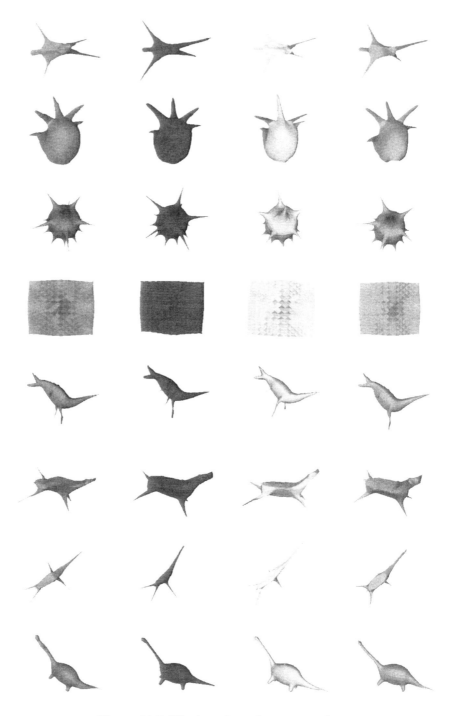

Figure 11.7: The invariant signature surfaces.

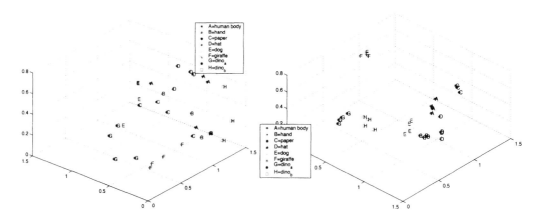

Figure 11.8: Applying moments based clustering to the original surfaces (left), and to their bending-invariant signature surfaces (right).

two weeks, Alex and Michael came up with a system that does indeed make this distinction. They got a straight A, and we immediately patented this important application.

In this case, one could implement directly the bending-invariant signatures method, and use it as a signature of the facial surface for recognition purposes. If we would also like to incorporate the surface color rather than just the geometric structure in the recognition process, then there are some alternatives for defining the facial surface and measuring the geodesic distance. One approach would be to define the face as a surface in a hybrid space of both space $\{X, Y, Z\}$, and color $\{R, G, B\}$, defined by its coordinates $\{X, Y, Z, R, G, B\}$. The facial surface, which is now described by both depth and color, can still be a 2D manifold, yet now embedded in $\mathbb{R}^6$. In this case, the arclength is defined by

$$ds^2 = ds_S^2 + \alpha ds_C^2,$$

where ds_S is an arclength on the manifold in the $\{X, Y, Z\}$ spatial subspace, and ds_C is an arclength in the $\{R, G, B\}$ color space. As in the case of the Beltrami framework, the aspect ratio constant α indicates the relative importance of color with respect to space. The fast marching method is applied as before, yet now the length of the triangle edges is measured in the hybrid space. A small α tells our flattening procedure to ignore the color, while a larger α increases the role of color in the flattening. In either cases, color can play a role in the recognition game, as the signature surface is isometric invariant in the $\{X, Y, Z\}$ space for any choice of α.

Figure 11.9 presents matching results of the reference subjects Michael 9 and Alex 20, using the eigenface method [120, 206] as shown at the top, and using the bending-invariant signatures in the middle row. We see how facial expressions influence the classification results of the eigenfaces method, unlike the bending-invariant signatures, which are insensitive to such deformations. Roughly speaking, the eigenface method takes all given face images in a training set, and generates a basis using principle component analysis by SVD. This procedure is also known as the Karhunen–Loève transform. The axes in this basis are ordered such that the first captures most of the average image energy—actually it is the average of all the images in the database. The second axis captures the second largest component, etc. Then, each new face image is projected onto this basis, and its first elements are used for recognition. Comparing between faces is thereby reduced to comparing the first few coefficients, instead of dealing with the whole image.

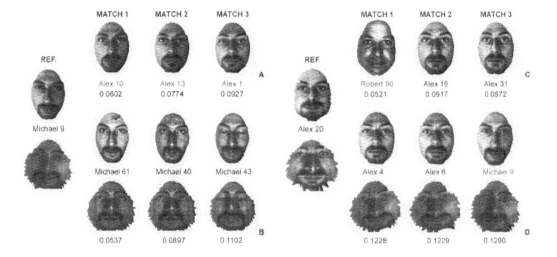

Figure 11.9: Comparing the isometric signature face matching (bottom) to the eigenface method (top).

The bending-invariant signatures outperform any method that uses only images and ignores the three-dimensional geometric structure of the face. Using this 3D face recognition by bending-invariant signatures, we actually overcome some of the main problems that arose in image-based biometrics. This is one example, which I hope the reader will find illuminating, of using geometric measures and numerical tools introduced in this book, to solve an interesting and challenging real world problem.

11.5 Exercises

1. Many reconstruction procedures in computer vision, such as photometric stereo, involve the reconstruction of a surface function $z(x, y)$ from its normals or gradient field ∇z. Show that ∇z is enough to compute the metric matrix $\mathbf{M} = (\delta_{ij}^2)$, and thereby, the bending-invariant signature. See [23].

2. Research project: Explore the problem of flat embedding a sphere in $\mathbb{R}^m$. Show that it is impossible to obtain Lipschitz flat embedding with vanishing error as m gets larger. Prove that the L_∞ error does not vanish for some specific cases. See [135]. Can the sphere be flattened for the L_2 error with a vanishing error as m increases?

3. Research project: Extend your results from Ex. 2 to more general surfaces.

Solutions to Selected Problems

Following are solutions to selected exercises numbered by the relevant chapter and the order of appearance of the exercise in that chapter.

- First part of the solution to Ex. 1.7 (Heron's formula)

  A simple proof for the area equation based on the law of cosines:

  $$\cos A = \frac{b^2 + c^2 - a^2}{2bc}$$

  and thus

  $$\sin A = \frac{-a^4 - b^4 - c^4 + 2b^2c^2 + 2c^2a^2 + 2a^2b^2}{2bc},$$

  which yields the known area equation

  $$
  \begin{aligned}
  \Delta &= \frac{1}{2}bc \sin A \\
  &= \frac{1}{4}\sqrt{-a^4 - b^4 - c^4 + 2b^2c^2 + 2c^2a^2 + 2a^2b^2} \\
  &= \frac{1}{4}\sqrt{(a+b+c)(-a+b+c)(a-b+c)(a+b-c)} \\
  &= \sqrt{s(s-a)(s-b)(s-c)}.
  \end{aligned}
  $$

- Solution to Ex. 1.8

  First, it is easy to prove that $\partial_t g(x,t) = \partial_{xx} g(x,t)$ for the Gaussian kernel g. Using this relation, we have for all t that

  $$
  \begin{aligned}
  \partial_t(g(x,t) * u_0) &= (\partial_t g(x,t)) * u_0 \\
  &= (\partial_{xx} g(x,t)) * u_0 \\
  &= \partial_{xx}(g(x,t) * u_0).
  \end{aligned}
  $$

Thus $u(t) = g(x, t) * u_0$ solves the heat equation. We add the relations $\lim_{t \to 0} g(x, t) = \delta(x)$, and $\delta(x) * u_0 = u_0$, and the recursive relation $g(x, 2dt) = g(x, dt) * g(x, dt)$, which integrate together to the desired proof.

- Solution to Ex. 1.10

The 2D kernel is given by the Gaussian

$$g(\bar{x}, t) = \frac{1}{4\pi t} e^{-\frac{\bar{x}^T \bar{x}}{4t}} = \frac{1}{4\pi t} e^{-\frac{x^2 + y^2}{4t}}.$$

The nD kernel is given by the nD normalized Gaussian

$$g(\bar{x}, t) = \frac{1}{(4\pi t)^{n/2}} e^{-\frac{\bar{x}^T \bar{x}}{4t}} = \frac{1}{4\pi t} e^{-\frac{x^2 + y^2}{4t}}.$$

- Solution to Ex. 1.11

The kernel is given by the affine Gaussian

$$g(\bar{x}, t) = \frac{1}{4\pi t \sqrt{\det(M)}} e^{-\frac{\bar{x}^T M^{-1} \bar{x}}{4t}},$$

where $\bar{x} = \begin{pmatrix} x \\ y \end{pmatrix}$.

- Solution to Ex. 2.2

$$
\begin{aligned}
C_v &= C_p P_v \\
C_{vv} &= (C_v)_v = (C_p P_v)_v = C_{pp} p_v^2 + C_p p_{vv} \\
1 = (C_v, C_{vv}) &= (C_p p_v, C_{pp} p_v^2 + C_p p_{vv}) \\
&= (C_p p_v, C_{pp} p_v^2) \\
&= p_v^3 (C_p, C_{pp})^{-1/3}.
\end{aligned}
$$

We conclude that $p_v = (C_p, C_{pp})^{-1/3}$. ∎

- Solution to Ex. 2.3

Start with the simplest area-preserving measure $(C_v, \hat{x}) = 1$. Differentiating this equation we get $(C_{vv}, \hat{x}) = 0$, that is $C_{vv} = \kappa_{shear} \hat{x}$. Prove that $v = \int (C_p, \hat{x}) dp$ is invariant under the uniform shear transformation. Next, show that $\langle C_{vv}, \vec{N} \rangle = (1 + \cot^2 \theta) \kappa = \kappa / \sin^2 \theta = \theta_s / \sin^2 \theta$, where $\theta = \angle(\hat{T}, \hat{x})$ is the angle between the tangent and the x-axis $\{1, 0\}$. The signature curve should be computed according to the invariant arclength $dv = |dy|$, and the shear curvature as the signature $\kappa_{shear} = \kappa |\sin \theta|^{-3}$.

- Solution to Ex. 2.7.

  Proof of Green, see Figure 12.1:

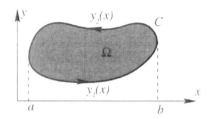

Figure 12.1: Proof of Green's theorem.

$$
\begin{aligned}
\iint_{\Omega} -f_y dx dy &= \int_a^b \int_{y_1(x)}^{y_2(x)} -f_y dy dx \\
&= -\int_a^b \left(f(x, y_2(x)) - f(x, y_1(x)) \right) dx \\
&= \int_a^b f(x, y_1(x)) dx - \int_a^b f(x, y_2(x)) dx \\
&= \int_a^b f(x, y_1(x)) dx + \int_b^a f(x, y_2(x)) dx \\
&= = \oint_C f dx,
\end{aligned}
$$

and similarly for g. For the area computation, set $g = x$ and $f = 0$, such that $g_x - f_y = 1$, and thereby

$$
\mathcal{A} \equiv \iint dx dy = \oint_C x dy = -\oint_C y dx;
$$

see Figure 12.2

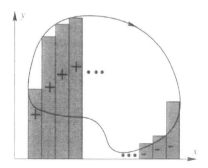

Figure 12.2: Area computation by Green's theorem.

- Solution to Ex. 3.4

Proof.

$$
\begin{aligned}
\frac{\partial}{\partial t}\frac{\partial}{\partial s} &= \frac{\partial}{\partial t}\left(\frac{1}{|C_p|}\frac{\partial}{\partial p}\right)\\
&= \left(\frac{\partial}{\partial t}\frac{1}{|C_p|}\right)\frac{\partial}{\partial p} + \frac{1}{|C_p|}\frac{\partial}{\partial t}\frac{\partial}{\partial p}\\
&= \left(\frac{\partial}{\partial t}\frac{1}{\langle C_p, C_p\rangle^{1/2}}\right)\frac{\partial}{\partial p} + \frac{1}{|C_p|}\frac{\partial}{\partial p}\frac{\partial}{\partial t}\\
&= -\frac{\langle C_{pt}, C_p\rangle}{|C_p|^3}\frac{\partial}{\partial p} + \frac{\partial}{\partial s}\frac{\partial}{\partial t}\\
&= -\frac{\langle \frac{\partial}{\partial p}V\vec{N}, \vec{T}\rangle}{|C_p|^2}\frac{\partial}{\partial p} + \frac{\partial}{\partial s}\frac{\partial}{\partial t}\\
&= -\frac{\langle V_p\vec{N} + V\vec{N}_p, \vec{T}\rangle}{|C_p|}\frac{\partial}{\partial s} + \frac{\partial}{\partial s}\frac{\partial}{\partial t}\\
&= -\langle V\vec{N}_s, \vec{T}\rangle\frac{\partial}{\partial s} + \frac{\partial}{\partial s}\frac{\partial}{\partial t}\\
&= -V\langle -\kappa\vec{T}, \vec{T}\rangle\frac{\partial}{\partial s} + \frac{\partial}{\partial s}\frac{\partial}{\partial t}\\
&= \frac{\partial}{\partial s}\frac{\partial}{\partial t} + V\kappa\frac{\partial}{\partial s}.
\end{aligned}
$$
∎

- Solution to Ex. 3.5

In general this is an open problem. Yet, assume you have numerical tools for measuring the area bounded by a closed curve and to position circles of minimal radius around any set of points in the plane. Then, a "morphological set-based" approach to establish a bound on the vanishing point location would be the following:

We know that a curve bounding an area of $A(0)$ will vanish at a point in $t_1 = A(0)/2\pi$. Find the circle with minimal radius that bounds the curve, say with radius r. This circle will vanish in $t_2 = \pi r^2/2\pi = r^2/2$. We know that the circle at t_1 (when the curve vanishes) is a circle with a radius r_1 that can be calculated:

$$
t_2 - t_1 = \frac{\pi r_1^2}{2\pi} = r_1^2/2,
$$

such that $r_1 = \sqrt{2t_2 - 2t_1} = \sqrt{r^2 - A(0)/\pi}$. We know that embedding is preserved along the evolution, and thus the curve should vanish inside its convex hull, and inside the eroded circle with r_1. Then, we can bound the vanishing point at the intersection of the convex hull and the eroded circle. See Figure 12.3. Note that the closer a curve is to a circle, the smaller the radius r_1.

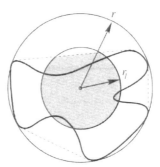

Figure 12.3: The vanishing point for a curve-shortening flow can be bounded by the intersection of the convex hull and a circle with radius $r_1 = \sqrt{r^2 - A(0)/\pi}$.

- Solution to Ex. 4.6

  a. The solution is $\phi_t = \beta(\kappa/(1 - \kappa\phi))$. This equation is used in [201], for better stability. **b.** The solution is $\phi_t = \beta((x, y) - \phi\nabla\phi)$. The vector $-\phi\nabla\phi$ maps the (x, y) point onto its closest corresponding point along the zero set. This equation is used in [81] for better efficiency and stability.

- Solution to Ex. 5.1

$$u_x = \frac{4u_{i+1} - 3u_i - u_{i+2}}{2\Delta x}.$$

- Solution to Ex. 5.4

  Let $\vec{f} = \{g, f\}$, then

$$\int_G \operatorname{div} \vec{f} \, dx dy = \int_G (g_x + f_y) dx dy = \oint_C (-f dx + g dy),$$

where C is the boundary of G and for the last relation we used Green's theorem. Recall that the normal to the boundary C can be written as $\vec{N} = C_s = \{x_s, y_s\} = \{-dy, dx\}/ds$, such that

$$\oint_C (-f dx + g dy) = -\oint_C \langle \vec{f}, \vec{N} \rangle ds,$$

- Solution to Ex. 6.1
  The solutions for a diamond and a square are

  $$\phi_t = |\phi_x| + |\phi_y|,$$
  $$\phi_t = \max\{|\phi_x|, |\phi_y|\}.$$

  For an ellipse the solution is a flow with a *hippopede*, also known as a *horse fetter*.

- Solution to Ex. 6.7
  It's the sets of points that bisects the lines.

- Solution to Ex. 9.2
  The EL is $\nabla g = C_{pp}$. Reparameterizing according to the arclength s yields eventually $p = s$, which should hold for the EL. For instance, $C_{ss} = \nabla g$, or equivalently $\nabla g = \kappa \vec{N}$. For the last equation to hold we need g to be constant along the curve C.

  Let us prove that last point. We have that the gradient of g should align with the curve normal, or in other words $\langle \nabla g, \vec{T} \rangle = 0$. But the inner product reads $g_x x_s + g_y y_s = g_s$ and by definition $g_s = 0$ implies that C is a level set of g, which proves the claim. ∎

- Solution to Ex. 9.6
  We prove that the steepest descent flow for the weighted area $\iint_\Omega g\,da$ is given by $C_t = g(x,y)\vec{N}$. Let the area enclosed by the contour be Ω. Next, define the functions $P(x,y)$ and $Q(x,y)$, such that $dP(x,y)/dx = g(x,y)/2$ and $dQ(x,y)/dy = g(x,y)/2$, and thus $g = P_x + Q_y$. Now, using Green's theorem we have

  $$\iint_\Omega g(x,y)\,dxdy = \iint_\Omega (P_x + Q_y)\,dxdy$$
  $$= \oint_0^L \langle \{P,Q\}, \vec{N} \rangle ds$$
  $$= \oint_0^1 \langle \{P,Q\}, \{-y_p, x_p\} \rangle dp$$
  $$= \oint_0^1 (-y_p P(C(p)) + x_p Q(C(p)))\,dp$$
  $$= \oint_0^1 F(C, C_p)\,dp,$$

  where $F(C, C_p) = -y_p P(C(p)) + x_p Q(C(p))$. The Euler–Lagrange is given by

  $$0 = \frac{dF}{dC} - \frac{d}{dp}\frac{\partial F}{\partial C_p}$$

$$
\begin{aligned}
&= -y_p \nabla P + x_p \nabla Q - \{\langle \nabla Q, C_p \rangle, -\langle \nabla P, C_p \rangle\} \\
&= \{-P_x y_p + Q_x x_p - Q_x x_p - Q_y y_p, -P_y y_p + Q_y x_p + P_x x_p \\
&\quad\; + P_y y_p\} \\
&= (P_x + Q_y)\{-y_p, x_p\} \\
&= g\vec{N},
\end{aligned}
$$

where in the last step we used our freedom of parameterization to obtain an intrinsic measure. ∎

- Solution to Ex. 9.7
  a and b. See Kimmel–Bruckstein Laplacian active contour paper [102].
  c. See Kimmel–Bruckstein paper that solves this problem [103].

- Solution to Ex. 9.11
  Proof. We first change the integration variable from an arclength s to an arbitrary parameter p.

$$
\begin{aligned}
E_A(C) = \int_0^L \langle \vec{V}, \vec{N} \rangle ds &= \int_0^L \langle \{u, v\}, \{-y_s, x_s\} \rangle ds \\
&= \int_0^1 \left\langle \{u, v\}, \frac{\{-y_p, x_p\}}{|C_p|} \right\rangle |C_p| dp \\
&= \int_0^1 (v x_p - u y_p) dp.
\end{aligned}
$$

Next, we compute the first variation for the x component,

$$
\begin{aligned}
\frac{\delta E_A(C)}{\delta x} &= \left(\frac{\partial}{\partial x} - \frac{d}{dp} \frac{\partial}{\partial x_p} \right)(v x_p - u y_p) \\
&= v_x x_p - u_x y_p - \frac{d}{dp} v \\
&= v_x x_p - u_x y_p - v_x x_p - v_y y_p \\
&= -y_p(u_x + v_y) = -y_p \mathrm{div}(\vec{V}).
\end{aligned}
$$

Similarly, for the y component we have $\delta E_A / \delta y = x_p \mathrm{div}(\vec{V})$. By freedom of parameterization, we end up with the first variation, $\delta E_A / \delta C = \mathrm{div}(\vec{V}) \vec{N}$. ∎

An important example is $\vec{V} = \nabla I$, for which we have as first variation

$$
\frac{\delta E_A(C)}{\delta C} = \Delta I \vec{N},
$$

where $\Delta I = I_{xx} + I_{yy}$ is the image Laplacian. The Euler–Lagrange equation $\delta E_A / \delta C = 0$ gives a variational explanation for the Marr–Hildreth edge detector that is defined by the zero crossings of the Laplacian, as first reported in [102].

- Solution to Ex. 9.12
  Actually the solution is identical to that of 9.7.
  Proof. We start by changing to an arbitrary parameter,

$$
E_{AR}(C) = \int_0^L |\langle \vec{V}, \vec{N} \rangle| ds \; = \; \int_0^1 \left| \left\langle \{v, u\}, \frac{\{-y_p, x_p\}}{|C_p|} \right\rangle \right| |C_p| dp
$$

$$
= \; \int_0^1 \sqrt{(ux_p - vy_p)^2} dp.
$$

Next, we compute the first variation for the x component,

$$
\frac{\delta E_{AR}(C)}{\delta x} \; = \; \left(\frac{\partial}{\partial x} - \frac{d}{dp} \frac{\partial}{\partial x_p} \right) \sqrt{(vx_p - uy_p)^2}
$$

$$
= \; -y_p \text{sign}(vx_p - uy_p)(u_x + v_y)
$$

$$
= \; -y_p \text{sign}(\langle \vec{V}, \vec{N} \rangle) \text{div}(\vec{V}).
$$

Similarly, for the y component we have

$$
\frac{\delta E_{AR}}{\delta y} \; = \; x_p \text{sign}(\langle \vec{V}, \vec{N} \rangle) \text{div}(\vec{V}).
$$

By freedom of parameterization, we end up with the first variation, $\delta E_{AR} / \delta C = \text{sign}(\langle \vec{V}, \vec{N} \rangle) \text{div}(\vec{V}) \vec{N}$. ∎

An important example is $\vec{V} = \nabla I$, for which we have

$$
\frac{\delta E_{AR}(C)}{\delta C} = \text{sign}(\langle \nabla I, \vec{N}(s) \rangle) \Delta I \vec{N}.
$$

- Solution to Ex. 9.13
  Proof. Following [220], we define the two functions $P(x, y)$ and $Q(x, y)$, such that $P_y(x, y) = -\frac{1}{2} f(x, y)$ and $Q_x = \frac{1}{2} f(x, y)$. We readily have that $f(x, y) = Q_x - P_y$. Next, using Green's theorem we can write

$$
E(C) = \iint_{\Omega_C} f(x, y) dx dy \; = \; \iint_{\Omega_C} (Q_x - P_y) dx dy
$$

$$
= \; \int_C (P dx + Q dy)
$$

$$
\begin{aligned}
&= \int_C (Px_s + Qy_s)ds \\
&= \int_C \langle \{-Q,P\}, \vec{N}\rangle ds,
\end{aligned}
$$

for which the first variation is given by Ex. 9.11, for $\vec{V} = \{-Q,P\}$, as

$$
\frac{\delta E(C)}{\delta C} = \mathrm{div}(\{-Q,P\})\vec{N} = -(Q_x - P_y)\vec{N} = -f\vec{N}. \qquad \blacksquare
$$

This term is sometimes called the weighted area [220] term, and for f constant, its resulting variation is known as the "balloon" [46] force. If we set $f = 1$, the gradient descent curve evolution process is the constant flow. It generates offset curves via $C_t = \vec{N}$, and its efficient implementation is closely related to Euclidean distance maps [55, 41] and fast marching methods [191].

- Solution to Ex. 9.14
  Proof.

$$
\begin{aligned}
E_{GAC}(C) = \int_0^L g(C(s))ds &= \int_0^1 g(C(p))|C_p|dp \\
&= \int_0^1 g(C(p))\sqrt{x_p^2 + y_p^2}\,dp.
\end{aligned}
$$

Next, we compute the first variation for the x component,

$$
\begin{aligned}
\frac{\delta E_{GAC}(C)}{\delta x} &= \left(\frac{\partial}{\partial x} - \frac{d}{dp}\frac{\partial}{\partial x_p} \right)\left(g(x(p),y(p))\sqrt{x_p^2 + y_p^2} \right) \\
&= g_x|C_p| - \frac{d}{dp}g\frac{x_p}{\sqrt{x_p^2 + y_p^2}} \\
&= g_x|C_p| - (g_x x_p + g_y y_p)\frac{x_p}{|C_p|} \\
&\quad - g\frac{x_{pp}|C_p| - x_p(x_p x_{pp} + y_p y_{pp})/|C_p|}{|C_p|^2} \\
&= y_p(\kappa g - \langle \nabla g, \vec{N}\rangle),
\end{aligned}
$$

where we used the curvature defined by

$$
\kappa = \frac{x_{pp}y_p - y_{pp}x_p}{|C_p|^3}.
$$

Similarly, for the y component we have $\delta E_{GAC}/\delta y = -x_p(\kappa g - \langle \nabla g, \vec{N}\rangle)$. By freedom of parameterization we end up with the above first variation. $\qquad \blacksquare$

We will use this term mainly for regularization. If we set $g = 1$, the gradient descent curve evolution result given by $C_t = -\delta E_{GAC}(C)/\delta C$ is the well-known curvature flow $C_t = \kappa \vec{N}$ or equivalently $C_t = C_{ss}$, also known as the geometric heat equation.

- Solution to Ex. 9.15

  Proof. Using Ex. 9.13, we have the first variation given by

  $$
  \begin{aligned}
  \frac{\delta E_{MV}}{\delta C} &= \frac{1}{2} \left((I - c_1)^2 - (I - c_2)^2 \right) \vec{N} \\
  &= \frac{1}{2} \left(I^2 - 2Ic_1 + c_1^2 - I^2 + 2Ic_2 - c_2^2 \right) \vec{N} \\
  &= \left((c_2 - c_1)I - \frac{(c_1 + c_2)(c_2 - c_1)}{2} \right) \vec{N} \\
  &= (c_2 - c_1) \left(I - \frac{c_1 + c_2}{2} \right) \vec{N}.
  \end{aligned}
  $$

  The optimal c_1 and c_2, extracted from $\delta E_{MV}/\delta c_1 = 0$ and $\delta E_{MV}/\delta c_2 = 0$, are the average intensities of the image inside and outside the contour, respectively. ∎

  One could recognize the variational interpretation of segmentation by the threshold $(c_1 + c_2)/2$ given by the Euler–Lagrange equation $\delta E_{MV}/\delta C = 0$.

- Solution to Ex. 9.16

  Proof. Using Ex. 9.13, we have the first variation $\delta E_{RMV}(C)/\delta C$.

  The optimal c_1 and c_2, extracted from $\delta E_{RMV}/\delta c_1 = 0$ and $\delta E_{RMV}/\delta c_2 = 0$, are the median intensities of the image inside and outside the contour, respectively:

  $$
  \begin{aligned}
  \frac{\delta E_{RMV}}{\delta c_1} &= \frac{d}{dc_1} \iint_{\Omega_C} \sqrt{(I - c_1)^2} dxdy \\
  &= -\iint_{\Omega_C} \frac{I - c_1}{|I - c_1|} dxdy \\
  &= -\iint_{\Omega_C} \operatorname{sign}(I - c_1) dxdy.
  \end{aligned}
  $$

  We have that $\iint_{\Omega_C} \operatorname{sign}(I - c_1) dxdy = 0$ for the selection of c_1 as the value of $I(x, y)$ in Ω_C that splits its area into two equal parts. From obvious reasons we refer to this value as the median of I in Ω_C, or formally,

  $$
  \begin{aligned}
  c_1 &= \operatorname{median}_{\Omega_C} I(x, y), \\
  c_2 &= \operatorname{median}_{\Omega \setminus \Omega_C} I(x, y).
  \end{aligned}
  $$

  ∎

The computation of c_1 and c_2 can be efficiently implemented via the intensity histograms in the interior or the exterior of the contour. One rough discrete approximation is the median of the pixels inside or outside the contour.

The robust minimal deviation term should be preferred when the penalty for isolated pixels with wrong affiliation is insignificant. The minimal variance measure penalizes large deviations in a quadratic fashion and would tend to oversegment such events or require large regularization that could oversmooth the desired boundaries.

- Solution to Ex. 9.17

```
%—————————————————————————————————
% Locally One-Dimensional (LOD) implicit scheme for
% closed geometric
% active contour model+minimal variation+GAC+robust alignment.
% I = Input image matrix,
% Phi = Initial contour matrix (implicit form).
% Balloon = Weighted area factor (scalar)
% Align = Alignment force factor (scalar)
% Max_Lloyd = The Max-Lloyd/Chan-Vese threshold factor (scalar)
% k = The time step (tau scalar)
% iter = Maximal number of iterations (scalar)
function Phi=LOD_Active_Contour(I,Phi,Balloon,...
                        Align,Max_Lloyd,k,iter)
D2I = Dxx(I)+Dyy(I); P = Dx(I); Q=Dy(I);
g = 1./sqrt(1+P.^2+Q.^2); % example for computing g(x,y)
delta = 2; count=1;
while and(delta >0.0001,count<iter) %check if converged
    Phi_old = Phi;
    threshold = LloydMax(Phi,I); % Max-Lloyd/Chan-Vese term
    alignment = -sign(P.*Dx(Phi)+Q.*Dy(Phi)).*D2I; % Laplacian
    Phi = Phi+k*(Balloon*g+Align*alignment+Max_Lloyd*threshold);
    for i=1:2, % i=1 => (I-tau*Ax) i=2 => (I-tau*Ay)
        Phi = Implicit(g(:),k,Phi); % (1/(I-tau*Ai))Phi
        Phi = Phi'; g = g'; % Transpose for Ay
    end % for i
    Phi = redistance(Phi); % Use fast marching for re-distancing
    delta = sum(sum((Phi_old-Phi).^2)) % Compute L2 norm
    count = count+1;
    imshow(I,[]); hold on; contour(Phi,[0 0],'r');hold off; drawnow;
end % while and function
%—————————————————————————————————
% Compute (1/(I- k*AI))Phi using Thomas algorithm
% k=time step, g=g(x,y) in column stack form
```

```
function u = Implicit(g,k,Phi)
gm = -k.*( g + g([end 1:end-1]))/2; % lower diag
gc = 1-k.*(-2*g - g([end 1:end-1]) -g([2:end 1]))/2; % main diag
gp = -k.*( g + g([2:end 1]))/2; % upper diag
u = Thomas(gc,gp(1:end-1),gm(2:end),Phi(:));
u = reshape(u,size(Phi));
%───────────────────────────────────────────
% Compute the Lloyd-Max/Chan-Vese thresholding
function force = LloydMax(Phi,I)
mask_in = (Phi<0); % inside the contour
mask_out = 1-mask_in; % rest of the domain
I_in = sum(sum(mask_in.*I))/sum(mask_in(:)); % mean value
I_out = sum(sum(mask_out.*I))/sum(mask_out(:));
force = (I_out-I_in).*(I-(I_in+I_out)/2);
%───────────────────────────────────────────
% 'Roughly' correct Phi to be a distance map of its zero set
function u=redistance(Phi);
u = (sign(Phi)+1)*999999; % set to infinity all positive
for i=1:2,
    I2 = 2;
    if i>1 u=(1-sign(Phi))*999999; end % set to infinity all negative
    while I2 > 1,
        v = Update(u,1);
        I2 = sum(sum((u-v).^2));
        u = v;
    end % while
    if i>1 u=up-u; else up=u; end %if
end % for
%───────────────────────────────────────────
% Solve |grad u|=F(x,y) parallel version of the FMM
function res = Update(u,F)
mx = min(u([2:end end],:), u([1 1:end-1],:));
my = min(u(:,[2:end end]), u(:,[1 1:end-1]));
delm = (mx -my);
mask = (abs(delm) < F);
res= min(mask.* (mx+my+sqrt(2.*F.^2- delm .^2))./2 + ...
                    (1-mask).* (min(mx,my)+F) , u);
%───────────────────────────────────────────
function f = Dmx(P)
f = P - P([1 1:end-1],:);
%───────────────────────────────
function f = Dpx(P)
f = P([2:end end],:) - P;
%───────────────────────────────
```

```
function f = Dx(P)
f = (Dpx(P)+Dmx(P))/2;
%————————————————
function f = Dy(P)
f = (Dx(P'))';
%————————————————
function f = Dxx(P)
f = Dpx(P)-Dmx(P);
%————————————————
function f = Dyy(P)
f = (Dxx(P'))';
%————————————————————————————
% Thomas Algorithm for trilinear diagonally dominant system:
% B u = d (solve for u); B is given by its 3 diagonals:
% alpha(1:N)=main diagonal, beta(1:N-1)=upper diagonal,
% gamma(1:N-1) = lower diagonal, (Compile first!)
function u = Thomas(alpha,beta,gamma,d)
N = length(alpha); r = beta;
l = zeros(N-1,1); u = zeros(N,1);
m = u; % zero
m(1) = alpha(1);
for i=1:N-1,
    l(i) = gamma(i)/m(i);
    m(i+1) = alpha(i+1)-l(i)*beta(i);
end % for
y = u; % zero
y(1) = d(1);
for i = 2:N,
    y(i) = d(i)-l(i-1)*y(i-1);
end % for
u(N) = y(N)/m(N);
for i = N-1:-1:1,
    u(i) = (y(i)-beta(i)*u(i+1))/m(i);
end % for
```

Bibliography

[1] D. Adalsteinsson and J. A. Sethian. A fast level set method for propagating interfaces. *J. of Comp. Phys.*, 118:269–277, 1995. [67, 133, 134]

[2] N. I. Akhiezer. *The calculus of variations.* Harwood Academic Publishers, 1988. Translated from Russian by Michael E. Alferieff. [26]

[3] L. Alvarez, F. Guichard, P. L. Lions, and J. M. Morel. Axioms and fundamental equations of image processing. *Arch. Rational Mechanics*, 123, 1993. [8, 11, 56, 77]

[4] L. Alvarez, P. L. Lions, and J. M. Morel. Image selective smoothing and edge detection by nonlinear diffusion. *SIAM J. Numer. Anal*, 29:845–866, 1992. [11]

[5] L. Alvarez and L. Mazorra. Signal and image restoration using shock filters and anisotropic diffusion. *SIAM J. Numer. Anal*, 31:590–605, 1994. [159]

[6] G. Aubert, P. Kornprobst. *Mathematical Problems in Image Processing: Partial Differential Equations and the Calculus of Variations.* Springer-Verlag, New York, 2002. [4]

[7] C. Ballester. An affine invariant model for image segmentation: Mathematical analysis and applications. Ph.D. thesis, Univ. Illes Balears, Palma de Mallorca, Spain, March 1995. [20]

[8] C. Ballester, V. Caselles, and M. Gonzalez. Affine invariant segmentation by variational method. Internal report, Univ. Illes Balears, Palma de Mallorca, Spain, 1995. [20]

[9] G. Barles and P. E. Souganidis. Convergence of approximation schemes for fully nonlinear second order equations. *Asymptotic Analysis*, 4:271–283, 1991. [73]

[10] T. Barth and J. A. Sethian. Numerical schemes for the Hamilton-Jacobi and level set equations on triangulated domains. *J. Computational Physics*, 145(1):1–40, 1998. [98]

[11] R. Bellman and R. Kalaba. *Dynamic Programming and Modern Control Theory*. London mathematical society monographs, London, 1965. [92]

[12] M. Bertalmo, L. T. Cheng, S. Osher, and G. Sapiro. Variational problems and partial differential equations on implicit surfaces: The framework and examples in image processing and pattern formation. *UMN-TR*, page www.ima.umn.edu/preprints/jun2000/jun2000.html, 2000. [59]

[13] M. Bichsel and A. P. Pentland. A simple algorithm for shape from shading. In *Proc. IEEE CVPR*, pp. 459–465, Champaign, IL, May 1992. [109, 122]

[14] A. Blake and A. Zisserman. *Visual Reconstruction*. MIT Press, Cambridge, MA, 1987. [143, 145]

[15] P. Blomgren and T. F. Chan. Color TV: Total variation methods for restoration of vector valued images. Cam TR, UCLA, 1996. [155]

[16] H. Blum. A transformation for extracting new descriptors of shape. In Walthen Dunn, editor, *Models for the Perception of Speech and Visual Form*, pp. 362–380. MIT Press, Cambridge, MA, 1967. [82]

[17] H. Blum. Biological shape and visual science (part I). *J. Theor. Biol.*, 38:205–287, 1973. [8, 82]

[18] I. Borg and P. Groenen. *Modern Multidimensional Scaling: Theory and Applications*. Springer, New York, 1997. [165]

[19] A. C. Bovik, M. Clark, and W. S. Geisler. Multichannel texture analysis using localized spatial filters. *IEEE Trans. on PAMI*, 12(1):55–73, 1990. [145]

[20] J. W. Brandt and V. R. Algazi. Continuous skeleton computation by voronoi diagram. *CVGIP: Image Understanding*, 55:329–338, 1992. [83]

[21] R. W. Brockett and P. Maragos. Evolution equations for continuous–scale morphology. In *Proc. IEEE Int. Conf. on Acoustics, Speech, and Signal Processing*, pp. 1–4, San Francisco, CA, March 1992. [8, 66, 75, 92]

[22] A. M. Bronstein, M. M. Bronstein, and R. Kimmel. Expression-invariant 3D face recognition. In *Proc. AVBPA, 4th Intl. Conf. on Audio- and Audio-based Biometric Person Authentication*, pp. 62–69 Surrey, UK, June 2003. [163, 164, 165, 173]

[23] A. M. Bronstein, M. M. Bronstein, R. Kimmel and A. Spira. 3D face recognition without facial surface reconstruction. Center for Intelligent Systems Report CIS #2003-05, Technion—Israel Institute of Technology, July 2003. [178]

[24] M. J. Brooks and W. Chojnacki. Direct computation of shape from shading. In *Proc. of ICPR, Intl. Conf. of Pattern Recognition*, pp. 114–119, Jerusalem, Israel, Oct. 1994. [110, 121]

[25] A. M. Bruckstein. On shape from shading. *Comput. Vision Graphics Image Process.*, 44:139–154, 1988. [4, 5, 109, 113]

[26] A. M. Bruckstein. Analyzing and synthesizing images by evolving curves. In *Proc. IEEE ICIP*, Vol. 1:11–15, Austin, Tx, Nov. 1994. [4]

[27] A. M. Bruckstein, R. J. Holt, A. Netravali, and T. J. Richardson. Invariant signatures for planar shape recognition under partial occlusion. *CVGIP: Image Understanding.*, 58(1):49–65, 1993. [20, 38]

[28] A. M. Bruckstein, N. Katzir, M. Lindenbaum, and M. Porat. Similarity invariant recognition of partially occluded planar curves and shapes. *International J. Computer Vision*, 7:271–285, 1992. [20]

[29] A. M. Bruckstein and A. Netravali. On differential invariants of planar curves and recognizing partially occluded planar shapes. AT&T technical report, AT&T, 1990. [20]

[30] A. M. Bruckstein and D. Shaked. On projective invariant smoothing and evolutions of planar curves and polygons. Center for Intelligent Systems Report CIS #9328, Technion—Israel Institute of Technology, Israel, Nov. 1993. [38]

[31] Su Buchin. *Affine Differential Geometry*. Science Press, Beijing, China, 1983. [22, 25]

[32] E. Calabi, P. J. Olver, and A. Tannenbaum. Affine geometry, curve flows, and invariant numerical approximations. Research report—The Geometry Center, University of Minnesota, June 1995. [22]

[33] V. Caselles, F. Catte, T. Coll, and F. Dibos. A geometric model for active contours. *Numerische Mathematik*, 66:1–31, 1993. [11, 124]

[34] V. Caselles, R. Kimmel, and G. Sapiro. Geodesic active contours. In *Proc. ICCV'95*, pp. 694–699, Boston, MA, June 1995. [11, 124, 145]

[35] V. Caselles, R. Kimmel, and G. Sapiro. Geodesic active contours. *International J. Computer Vision*, 22(1):61–79, 1997. [11, 124, 134]

[36] V. Caselles, R. Kimmel, G. Sapiro, and C. Sbert. Minimal surfaces: A geometric three dimensional segmentation approach. *Numerische Mathematik*, 77(4):423–425, 1997. [45]

[37] V. Caselles, R. Kimmel, G. Sapiro, and C. Sbert. Minimal surfaces based object segmentation. *IEEE Trans. on PAMI*, 19:394–398, 1997. [45, 132]

[38] A. Chambolle. Partial differential equations and image processing. In *Proc. IEEE ICIP*, Vol. 1:16–20, Austin, TX, Nov. 1994. [145, 155, 156]

[39] T. Chan and L. Vese. An active contour model without edges. In *Scale-Space Theories in Computer Vision*, pp. 141–151, 1999. [134, 139]

[40] T. F. Chan, G. H. Golub, and P. Mulet. A nonlinear primal-dual method for total variation-based image restoration. *Presented at AMS/SIAM Workshop on Linear and Nonlinear CG Methods*, July 1995. [46]

[41] C. S. Chiang, C. M. Hoffmann, and R. E. Lync. How to compute offsets without self-intersection. In *Proc. of SPIE*, Vol. 1620: 76, 1992. [134, 187]

[42] D. L. Chopp. Computing minimal surfaces via level set curvature flow. Ph.D Thesis, Lawrence Berkeley Lab. and Dept. of Math. LBL-30685, Univ. of CA, Berkeley, May 1991. [47, 134]

[43] D. L. Chopp. Computing minimal surfaces via level set curvature flow. *J. of Computational Physics*, 106(1):77–91, May 1993. [46, 47, 57, 67, 133]

[44] D. L. Chopp and J. A. Sethian. Flow under curvature: Singularity formation, minimal surfaces, and geodesics. *J. Exper. Math.*, 2(4):235–255, 1993. [46, 57, 133]

[45] I. Cohen and L. D. Cohen. Finite element methods for active contour models and balloons for 2-D and 3-D images. *IEEE Trans. on PAMI*, 15(11), 1993. [126]

[46] L. D. Cohen. On active contour models and balloons. *CVGIP: Image Understanding*, 53(2):211–218, 1991. [126, 187]

[47] L. D. Cohen and R. Kimmel. Global minimum for active contours models: A minimal path approach. *International J. Computer Vision*, 24(1):57–78, 1997. [11, 130, 138]

[48] T. Cohignac, L. Lopez, and J. M. Morel. Integral and local affine invariant parameters and application to shape recognition. In D. Dori and A. M. Bruckstein, editors, *Shape Structure and Pattern Recognition*. World Scientific Publishing, Singapore, 1995. [20]

[49] P. Concus. Numerical solution of the minimal surface equation. *Mathematics of Computation*, 21:340–350, 1967. [46]

[50] G. H. Cottet and L. Germain. Image processing through reaction combined with nonlinear diffusion. *Math. Comp.*, 61:659–673, 1993. [156]

[51] R. Courant, K. O. Friedrichs, and H. Lewy. Uber die partiellen Differenzengliechungen der mathematisches Physik. *Math. Ann.*, 100:32–74, 1928. [67]

[52] R. Courant, K. O. Friedrichs, and H. Lewy. On the partial difference equations of mathematical physics. *IBM Journal*, 11:215–235, 1967. [67]

[53] M. A. A. Cox and T. F. Cox. *Multidimensional Scaling*. Chapman and Hall, London, 1994. [165]

[54] M. G. Crandall, H. Ishii, and P. L. Lions. User's guide to viscosity solutions of second order partial linear differential equations. *Bull. American Math. Society*, 27:1–67, 1992. [72, 109]

[55] P. Danielsson. Euclidean distance mapping. *Computer Graphics and Image Processing*, 14:227–248, 1980. [80, 90, 187]

[56] S. Di Zenzo. A note on the gradient of a multi image. *Computer Vision, Graphics, and Image Processing*, 33:116–125, 1986. [156]

[57] E. W. Dijkstra. A note on two problems in connection with graphs. *Numerische Mathematic*, 1:269–271, 1959. [94, 95]

[58] M. P. DoCarmo. *Differential Geometry of Curves and Surfaces*. Prentice-Hall Inc., Englewood Cliffs, NJ, 1976. [108, 169]

[59] L. Dorst and R. Boomgaard. Morphological signal processing and the slope transform. *Signal Processing*, 38:79–98, 1994. [75]

[60] E. R. Dougherty. *An Introduction to Morphological Image Processing*. SPIE Press, Bellingham, WA, 1984. [75]

[61] P. Dupuis and J. Oliensis. Direct method for reconstructing shape from shading. In *Proc. IEEE CVPR*, pp. 453–458, Champaign, IL, May 1992. [109]

[62] P. Dupuis and J. Oliensis. An optimal control formulation and related numerical methods for a problem in shape reconstruction. *Ann. Appl. Probab.*, 4(2):287–346, 1994. [121, 122]

[63] A. Elad and R. Kimmel. Bending invariant representations for surfaces. In *Proc. IEEE CVPR, Computer Vision and Pattern Recognition*, pp. 162–167, Hawaii, Dec. 2001. [164, 166, 171]

[64] A. Elad and R. Kimmel. On bending invariant signatures for surfaces. *IEEE Trans. on Pattern Analysis and Machine Intelligence*, to appear, 2003 [164, 166, 171]

[65] A. I. El-Fallah, G. E. Ford, V. R. Algazi, and R. R. Estes. The invariance of edges and corners under mean curvature diffusions of images. In *Processing III SPIE*, Vol. 2421:2–14, 1994. [145]

[66] G. Elber and E. Cohen. Error bounded variable distance offset operator for free form curves and surfaces. *Intl. J. Comput. Geom. & Applic.*, 1(1):67–78, 1991. [106]

[67] C. L. Epstein and M. Gage. The curve shortening flow. In A. Chorin and A. Majda, editors, *Wave Motion: Theory, Modeling, and Computation*. Springer-Verlag, New York, 1987. [37, 38]

[68] L. C. Evans and J. Spruck. Motion of level sets by mean curvature, I. *J. Diff. Geom.*, 33, 1991. [56]

[69] O. Faugeras and R. Keriven. Variational principles, surface evolution PFE's, level set methods, and the stereo problem. *IEEE Trans. on Image Processing*, 7(3):336–344, 1998. [133]

[70] M. S. Floater. Parameterization and smooth approximation of surface triangulations. *Comp. Aided Geom. Design* 14:231–250, 1977. [169]

[71] M. S. Floater. Parametric tilings and scattered data approximation. *International J. of Shape Modeling*, 4:165–182, 1998. [169]

[72] S. Fortune. A sweepline algorithm for voronoi diagrams. *Algorithmica*, 2:153–174, 1987. [105]

[73] D. Gabor. Information theory in electron microscopy. *Laboratory Investigation*, 14(6):801–807, 1965. [144, 156, 158, 159]

[74] M. Gage and R. S. Hamilton. The heat equation shrinking convex plane curves. *J. Diff. Geom.*, 23, 1986. [11, 38, 44, 48]

[75] I. M. Gelfand and S. V. Fomin. *Calulus of Variations*. Prentice-Hall, Inc., Englewood Cliffs, NJ, 1963. [24]

[76] J. Gil and R. Kimmel. Efficient dilation, erosion, opening and closing algorithms. *IEEE Trans. on Pattern Analysis and Machine Intelligence*, 24(12):1606–1617, 2002. [75]

[77] J. Gil and M. Werman. Computing 2-D min, median, and max filters. *IEEE Trans. on Pattern Analysis and Machine Intelligence*, 15(5):504–507, May 1993. [75]

[78] S. K. Godunov. Finite difference method for numerical computation of discontinuous solution of the equations of fluid dynamics. *Matematicheskii Sbornik*, 47:271, 1959. Translated from Russian by I. Bohachevsky. [92]

[79] R. Goldenberg, R. Kimmel, E. Rivlin, and M. Rudzsky. Fast geodesic active contours. *IEEE Trans. on Image Processing*, 10(10):1467–1475, 2001. [134]

[80] R. Goldenberg, R. Kimmel, E. Rivlin, and M. Rudzsky. Fast geodesic active contours. In *Lecture Notes in CS: Scale-Space Theories in Computer Vision*, Vol. 1682, Springer, New York, 1999. [133]

[81] J. Gomes and O. Faugeras. Reconciling distance functions and level sets. In *Lecture Notes in CS: Scale-Space Theories in Computer Vision*, Vol. 1682. Springer, New York, 1999. [183]

[82] M. A. Grayson. The heat equation shrinks embedded plane curves to round points. *J. Diff. Geom.*, 26, 1987. [11, 44, 56]

[83] M. A. Grayson. Shortening embedded curves. *Annals of Mathematics*, 129:71–111, 1989. [57]

[84] R. Grossman, N. Kiryati, and R. Kimmel. Computational surface flattening: A voxel-based approach. *IEEE Trans. on Pattern Analysis and Machine Intelligence*, 24(4):433–441, 2002. [164]

[85] F. Guichard and J. M. Morel. *Image iterative smoothing and P.D.E.'s*. Online Book, Sept. 1998. [49]

[86] A. Harten, J. M. Hyman, and P. D. Lax. On finite-difference approximations and entropy conditions for shocks. *Comm. Pure Appl. Math.*, 29:297, 1976. [65]

[87] B. K. P. Horn. Obtaining shape from shading information. In P. H. Winston, editor, *The Psychology of Computer Vision*, pp. 115–155. McGraw Hill, New York, 1975. [5, 109]

[88] B. K. P. Horn. Height and gradient from shading. *International J. Computer Vision*, 5:37–75, 1990. [109]

[89] B. K. P. Horn and M. J. Brooks, editors. *Shape from Shading*. MIT Press, Cambridge, MA, 1989. [5, 109]

[90] M. Kass, A. Witkin, and D. Terzopoulos. Snakes: Active contour models. *International J. Computer Vision*, 1:321–331, 1988. [4, 123]

[91] S. Kichenassamy, A. Kumar, P. Olver, A. Tannenbaum, and A. Yezzi. Gradient flows and geometric active contour models. In *Proc. ICCV'95*, pp. 810–815, Boston, MA, June 1995. [125]

[92] R. Kimmel. Shape from shading via level sets. M.Sc. thesis (in Hebrew), Technion—Israel Institute of Technology, June 1992. [5, 115]

[93] R. Kimmel. *Curve Evolution on Surfaces*. D.Sc. thesis, Technion—Israel Institute of Technology, June 1995. [4]

[94] R. Kimmel. Intrinsic scale space for images on surfaces: The geodesic curvature flow. *Graphics Modeling and Image Processing*, 59(5):365–372, Sept. 1997. [57]

[95] R. Kimmel. 3D shape reconstruction from autostereograms and stereo. *Special issue on PDEs in Image Processing, Computer Vision, and Computer Graphics, Journal of Visual Communication and Image Representation*, 13:324–333, 2002. [133]

[96] R. Kimmel. Fast edge integration. In *Level Set Methods and Their Applications in Computer Vision*, Ch. 3. Springer-Verlag, New York, 2002. [133]

[97] R. Kimmel, A. Amir, and A. M. Bruckstein. Finding shortest paths on surfaces using level sets propagation. *IEEE Trans. on PAMI*, 17(6):635–640, June 1995. [9, 95, 106]

[98] R. Kimmel and A. M. Bruckstein. Shape from shading via level sets. Center for Intelligent Systems Report CIS #9209, Technion—Israel Institute of Technology, Israel, June 1992. [109, 115]

[99] R. Kimmel and A. M. Bruckstein. Shape offsets via level sets. *CAD*, 25(5):154–162, 1993. [8, 48, 106]

[100] R. Kimmel and A. M. Bruckstein. Global shape from shading. *CVIU*, 62(3):360–369, Nov. 1995. [7, 110, 116, 118, 121]

[101] R. Kimmel and A. M. Bruckstein. Tracking level sets by level sets: A method for solving the shape from shading problem. *CVIU*, 62(2):47–58, July 1995. [5, 109, 115]

[102] R. Kimmel and A. M. Bruckstein. Regularized Laplacian zero crossings as optimal edge integrators. In *Proc. Image and Vision Computing, IVCNZ01*, New Zealand, Nov. 2001. [134, 185, 186]

[103] R. Kimmel and A. M. Bruckstein. On edge detection edge integration and geometric active contours. In *Proc. Intl. Symposium on Mathematical Morphology, ISMM2002*, Sydney, New South Wales, Australia, April 2002. [134, 185]

[104] R. Kimmel and N. Kiryati. Finding shortest paths on surfaces by fast global approximation and precise local refinement. *Intl. J. of Pattern Rec. and AI*, 10(6):643–656, 1996. [57]

[105] R. Kimmel, N. Kiryati, and A. M. Bruckstein. Sub-pixel distance maps and weighted distance transforms. *J. Mathematical Imaging and Vision, Special Issue on Topology and Geometry in Computer Vision*, 6:223–233, June 1996. [10, 106]

[106] R. Kimmel, N. Kiryati, and A. M. Bruckstein. Multi-valued distance maps in finding shortest paths between moving obstacles. *IEEE Trans. on Robotics and Automation*, 14(3):427–436, 1998. [95]

[107] R. Kimmel, R. Malladi, and N. Sochen. Image processing via the beltrami operator. In *Proc. of 3rd Asian Conf. on Computer Vision*, pp. 574–581, Hong Kong, Jan. 1998. Springer-Verlag, LNCS 1351. [55, 144, 161]

[108] R. Kimmel, R. Malladi, and N. Sochen. Images as embedding maps and minimal surfaces: Movies, color, texture, and volumetric medical images. *International J. Computer Vision*, 39(2):111–129, 2000. [59, 144, 150, 161]

[109] R. Kimmel and G. Sapiro. Shortening three dimensional curves via two dimensional flows. *International Journal: Computers & Mathematics with Applications*, 29(3):49–62, 1995. [9, 57]

[110] R. Kimmel and G. Sapiro. The mathematics of face recognition. *SIAM News*, 36(3), April 2003 [163]

[111] R. Kimmel and J. A. Sethian. Optimal algorithm for shape from shading and path planning. *J. Mathematical Imaging and Vision*, 14(3):237–244, 2001. [95]

[112] R. Kimmel and J. A. Sethian. Computing geodesic paths on manifolds. *Proc. of National Academy of Sciences, USA*, 95(15):8431–8435, 1998. [9, 87, 98, 102]

[113] R. Kimmel and J. A. Sethian. An optimal time algorithm for shape from shading. Report LBNL-41660, Berkeley Labs, UC, CA 94720, April 1998. [118, 121]

[114] R. Kimmel and J. A. Sethian. Fast voronoi diagrams and offsets on triangulated surface. In *Curves and Surfaces*. A. K. Peters, Ltd., Wellesley, MA, 1999. [102, 105]

[115] R. Kimmel, D. Shaked, N. Kiryati, and A. M. Bruckstein. Skeletonization via distance maps and level sets. *CVIU*, 62(3):382–391, Nov. 1995. [10, 82, 84]

[116] R. Kimmel, K. Siddiqi, B. B. Kimia, and A. M. Bruckstein. Shape from shading: Level set propagation and viscosity solutions. *International J. Computer Vision*, 16:107–133, 1995. [5, 115]

[117] R. Kimmel, N. Sochen, and R. Malladi. On the geometry of texture. Report LBNL-39640, UC-405, Berkeley Labs, UC, CA 94720, Nov. 1996. [55, 145]

[118] R. Kimmel, N. Sochen, and R. Malladi. From high energy physics to low level vision. In *Lecture Notes in Computer Science: First International Conference on Scale-Space Theory in Computer Vision*, Vol. 1252:236–247. Springer-Verlag, New York, 1997. [55, 144, 149, 150]

[119] R. Kimmel, N. Sochen, and R. Malladi. Images as embedding maps and minimal surfaces: Movies, color, and volumetric medical images. In *Proc. of IEEE CVPR'97*, pp. 350–355, Puerto Rico, June 1997. [144]

[120] M. Kirby and L. Sirovich. Application of the Karhunen–Loeve procedure for the characterization of human faces. *IEEE Trans. on Pattern Analysis and Machine Intelligence*, 12(1):103–108, 1990. [176]

[121] N. Kiryati and G. Székely. Estimating shortest paths and minimal distances on digitized three dimensional surfaces. *Pattern Recognition*, 26(11):1623–1637, 1993. [10]

[122] E. Kreyszig. *Differential Geometry*. Dover Publications, Inc., New York, 1991. [171]

[123] J. B. Kruskal and M. Wish. *Multidimensional Scaling*. Sage Publications, Thousand Oaks, CA, 1978. [165]

[124] R. Kunze, F. E. Wolter, and T. Rausch. Geodesic Voronoi diagrams on parametric surfaces. Welfen lab. report, Welfen Lab., Univ. Hannover, Germany, July 1997. [106]

[125] N. N. Kuznetsov. Accuracy of some approximate methods for computing the weak solutions of first-order quasi-linear equation. *USSR Comp. Math. and Math Phys.*, 16:105–119, 1976. [65]

[126] J. L. Lagrange. *Essai d'une nouvelle méthode pour déterminer les maxima et les minima des formules intégrales indéfinies*, Vol. I. Gauthier-Villars, Paris, 1867. Translated by D. J. Struick. [45]

[127] J. C. Latombe. *Robot Motion Planning*. Kluwer Academic Publishers, Boston, MA, 1991. [95]

[128] P. D. Lax. *Hyperbolic Systems of Conservation Laws and the Mathematical Theory of Shock Waves*. Society for Industrial and Applied Mathematics, Philadelphia, PA, 1973. [63]

[129] C. H. Lee and A. Rosenfeld. Improved methods of estimating shape from shading using the light source coordinate system. *Artificial Intelligence*, 26(1):125–143, May 1985. [119, 120]

[130] D. T. Lee. Medial axis transformation of a planar shape. *IEEE Trans. on PAMI*, 4:363–369, 1982. [83]

[131] T. S. Lee. Image representation using 2D Gabor-wavelets. *IEEE Trans. on PAMI*, 18(10):959–971, 1996. [145]

[132] T. S. Lee, D. Mumford, and A. L. Yuille. Texture segmentation by minimizing vector valued energy functionals: The couple-membrane model. In G. Sandini, editor, *Lecture Notes in Computer Science, 588, Computer Vision: ECCV'92*, pp. 165–173. Springer-Verlag, New York, 1992. [145]

[133] R. J. LeVeque. *Numerical Methods for Conservation Laws*. Lectures in Mathematics. Birkhauser Verlag, Basel, 1992. [67]

[134] M. Lindenbaum, M. Fischer, and A. M. Bruckstein. On Gabor's contribution to image enhancement. *Pattern Recognition*, 27(1):1–8, 1994. [156, 158]

[135] N. Linial. Finite metric spaces—combinatorics, geometry and algorithms. In *Proc. Int. Congress of Mathematicians III*, pp. 573–586, Beijing, 2002. [178]

[136] P. L. Lions, E. Rouy, and A. Tourin. Shape-from-shading, viscosity solutions and edges. *Numerische Mathematik*, 64:323–353, 1993. [121]

[137] T. Lu, P. Neittaanmaki, and X-C. Tai. A parallel splitting up method and its application to Navier-Stokes equations. *Applied Mathematics Letters*, 4(2):25–29, 1991. [133, 134]

[138] T. Lu, P. Neittaanmaki, and X-C. Tai. A parallel splitting up method for partial differential equations and its application to Navier-Stokes equations. *RAIRO Math. Model. and Numer. Anal.*, 26(6):673–708, 1992. [133, 134]

[139] D. L. MacAdam. Visual sensitivity to color differences in daylight. *J. Opt. Soc. Am.*, 32:247, 1942. [151]

[140] D. L. MacAdam. Specification of small chromaticity differences. *J. Opt. Soc. Am.*, 33:18, 1943. [151]

[141] T. Maekawa. Computation of shortest paths on free-form parametric surfaces. *J. Mechanical Design*, 118:499–508, 1996. [106]

[142] R. Malladi and J. A. Sethian. Image processing: Flows under min/max curvature and mean curvature. *Graphical Models and Image Processing*, 58(2):127–141, March 1996. [145]

[143] R. Malladi and J. A. Sethian. An O(N log N) algorithm for shape modeling. *Proc. of National Academy of Sciences, USA*, 93:9389–9392, 1996. [132]

[144] R. Malladi, J. A. Sethian, and B. C. Vemuri. A topology-independent shape modeling scheme. In *SPIE's Geometric Methods in Computer Vision II*, Vol. SPIE 2031:246–255, July 1993. [124]

[145] R. Malladi, J. A. Sethian, and B. C. Vemuri. Evolutionary fronts for topology-independent shape modeling and recovery. In *Proc. Third European Conf. on Computer Vision*, pp. 3–13, Stockholm, Sweden, May 1994. [11, 124]

[146] R. Malladi, J. A. Sethian, and B. C. Vemuri. Shape modeling with front propagation: A level set approach. *IEEE Trans. on PAMI*, 17:158–175, 1995. [11, 124]

[147] B. S. Manjunath and W. Y. Ma. Texture features for browsing and retrieval of image data. *IEEE Trans. on PAMI*, 18(8):837–841, 1996. [145]

[148] P. Maragos. Slope transforms: Theory and application to nonlinear signal processing. *IEEE Trans. on Signal Processing*, 43(4):864–877, 1995. [75]

[149] J. S. B. Mitchell, D. M. Mount, and C. H. Papadimitriou. The discrete geodesic problem. *SIAM J. Comput.*, 16(4):647–668, 1987. [106]

[150] U. Montanari. A method for obtaining skeletons using a quasi-Euclidean distance. *J. Association for Computing Machinery*, 16(4):534–549, 1968. [82]

[151] U. Montanari. Continuous skeletons from digitized images. *J. Association for Computing Machinery*, 16(4):534–549, 1969. [82]

[152] T. Moons, E J. Pauwels, L. J. Van Gool, and A. Oosterlinck. Foundations of semi-differential invariants. *International J. Computer Vision*, 14(1):49–65, 1995. [20, 25, 38]

[153] D. Mumford and J. Shah. Boundary detection by minimizing functionals. In *Proc. CVPR, Computer Vision and Pattern Recognition*, pp. 22–26, San Francisco, 1985. [150]

[154] R. L. Ogniewicz. *Discrete Voronoi Skeletons*. Hartung–Gorre Verlag Konstanz, Zürich, 1993. [83]

[155] T. Ohta, D. Jansow, and K. Karasaki. Universal scaling in the motion of random interfaces. *Physical Review Letters*, 49(17):1223–1226, 1982. [50]

[156] P. J. Olver, G. Sapiro, and A. Tannenbaum. Invariant geometric evolutions of surfaces and volumetric smoothing. MIT report - LIDS, MIT, April 1994. [38]

[157] Peter J. Olver. *Equivalence, Invariants, and Symmetry*. Cambridge University Press, Cambridge, UK, 1995. [22]

[158] S. Osher and R. Fedkiw *Level Set Methods and Dynamic Implicit Surfaces* Springer-Verlag, New York, 2003 [7]

[159] S. J. Osher and C. W. Shu. High-order essentially nonoscillatory schemes for Hamilton–Jacobi equations. *SIAM J. Numer. Analy.*, 28(4):907–922, August 1991. [64, 66, 67]

[160] S. J. Osher and L. I. Rudin. Feature-oriented image enhancement using shock filters. *SIAM J. Numer. Analy.*, 27(4):919–940, August 1990. [159]

[161] S. J. Osher and J. A. Sethian. Fronts propagating with curvature dependent speed: Algorithms based on Hamilton–Jacobi formulations. *J. of Comp. Phys.*, 79:12–49, 1988. [4, 50, 51, 61, 66, 67, 106]

[162] N. Paragios and R. Deriche. A PDE-based level set approach for detection and tracking of moving objects. In *Proc. of the 6th ICCV*, Bombay, India, 1998. [133]

[163] N. Paragios and R. Deriche. Geodesic active contours and level sets for the detection and tracking of moving objects. *IEEE Trans. on PAMI*, 22(3):266–280, 2000. [134]

[164] P. Perona and J. Malik. Scale-space and edge detection using anisotropic diffusion. *IEEE-PAMI*, 12:629–639, 1990. [156]

[165] Y. Pnueli and A. M. Bruckstein. DigiDurer—a digital engraving system. *The Visual Computer*, 10:277–292, 1994. [5, 7]

[166] Y. Pnueli and A. M. Bruckstein. Gridless halftoning. *Graphic Models and Image Processing*, 58(1):38–64, 1996. [5, 7]

[167] M. Porat and Y. Y. Zeevi. The generalized Gabor scheme of image representation in biological and machine vision. *IEEE Trans. on PAMI*, 10(4):452–468, 1988. [145]

[168] In B. M. ter Haar Romeny, editor, *Geometric-Driven Diffusion in Computer Vision*. Kluwer Academic Publishers, The Netherlands, 1994. [150]

[169] M. Proesmans, E. Pauwels, and L. van Gool. Coupled geometry-driven diffusion equations for low level vision. In B. M. ter Haar Romeny, editor, *Geometric-Driven Diffusion in Computer Vision*, pp. 191–228. Kluwer Academic Publishers, The Netherlands, 1994. [156]

[170] E. Rouy and A. Tourin. A viscosity solutions approach to shape-from-shading. *SIAM. J. Numer. Analy.*, 29(3):867–884, June 1992. [73, 92, 109, 115]

[171] Y. Rubner and C. Tomasi. Coalescing texture descriptors. In *Proc. the ARPA Image Understanding Workshop*, pp. 927–936, Feb. 1996. [145]

[172] Y. Rubner and C. Tomasi. Texture metrics. In *Proc. IEEE Conf. on Systems, Man, and Cybernetics*, pp. 4601–4607, 1998. [165]

[173] L. Rudin, S. Osher, and E. Fatemi. Nonlinear total variation based noise removal algorithms. *Physica D*, 60:259–268, 1992. [127, 142, 150]

[174] G. Sapiro. *Topics in Shape Evolution*. D.Sc. thesis (in Hebrew), Technion— Israel Institute of Technology, April 1993. [25, 38, 44, 48]

[175] G. Sapiro. Vector-valued active contours. In *Proc. IEEE CVPR'96*, pp. 680–685, 1996. [145, 155]

[176] G. Sapiro. *Geometric Partial Differential Equations and Image Analysis*. Cambridge University Press, 2001 [4]

[177] G. Sapiro, R. Kimmel, D. Shaked, B. Kimia, and A. M. Bruckstein. Implementing continuous-scale morphology via curve evolution. *Pattern Recognition*, 26(9):1363–1372, 1993. [8, 75]

[178] G. Sapiro and D. Ringach. Anisotropic diffusion of multivalued images. In *12th Intl. Conf. on Analysis and Optimization of Systems: Images, Wavelets and PDE's*, Vol. 219 of *Lecture Notes in Control and Information Sciences*, pp. 134–140, Springer, London, 1996. [156]

[179] G. Sapiro and D. L. Ringach. Anisotropic diffusion of multivalued images with applications to color filtering. *IEEE Trans. Image Proc.*, 5:1582–1586, 1996. [145, 151, 155, 156]

[180] G. Sapiro and A. Tannenbaum. Affine invariant scale-space. *International J. Computer Vision*, 11(1):25–44, 1993. [11, 20, 56]

[181] G. Sapiro and A. Tannenbaum. On invariant curve evolution and image analysis. *Indiana University Mathematics J.*, 42(3), 1993. [23, 44]

[182] G. Sapiro and A. Tannenbaum. Area and length preserving geometric invariant scale-spaces. *IEEE Trans. on PAMI*, 17(1), 1995. [5]

[183] E. Schrödinger. Grundlinien einer theorie der farbenmetrik in tagessehen. *Ann. Physik*, 63:481, 1920. [150]

[184] Schroeder. The eikonal equation. *The Mathematical Intelligencer*, 5(1):36–37, 1983. [7]

[185] E. L. Schwartz, A. Shaw, and E. Wolfson. A numerical solution to the generalized map-maker's problem: Flattening non-convex polyhedral surfaces. *IEEE Trans. on Pattern Analysis and Machine Intelligence*, 11(9):1005–1008, 1989. [164, 166]

[186] R. Sedgewick. *Algorithms*. Addison-Wesley, Reading, MA, 1988. [93, 94]

[187] J. A. Sethian. Curvature and the evolution of fronts. *Commun. in Math. Phys.*, 101:487–499, 1985. [48, 51, 64]

[188] J. A. Sethian. Numerical methods for propagating fronts. In P. Concus and R. Finn, editors, *Variational Methods for Free Surface Interfaces*. Springer-Verlag, New York, 1987. [64]

[189] J. A. Sethian. A review of recent numerical algorithms for hypersurfaces moving with curvature dependent speed. *J. of Diff. Geom.*, 33:131–161, 1990. [50, 51, 53, 64]

[190] J. A. Sethian. A review of the theory, algorithms, and applications of level set methods for propagating interfaces. *Acta Numerica*, Cambridge University Press, 1995. [87, 94, 134]

[191] J. A. Sethian. *Level Set Methods: Evolving Interfaces in Geometry, Fluid Mechanics, Computer Vision and Materials Sciences*. Cambridge University Press, 1996. [87, 94, 106, 187]

[192] J. A. Sethian. A marching level set method for monotonically advancing fronts. *Proc. Nat. Acad. Sci.*, 93(4), 1996. [87, 94]

[193] J. Shah. A common framework for curve evolution, segmentation and anisotropic diffusion. In *Proc. IEEE CVPR'96*, pp. 136–142, 1996. [125]

[194] J. Shah. Curve evolution and segmentation functionals: Application to color images. In *Proc. IEEE ICIP'96*, pp. 461–464, 1996. [155]

[195] D. Shaked and A. M. Bruckstein. The curve axis. *CVGIP: Computer Vision and Image Understanding*, 63(2):367–379, 1996. [10]

[196] I. Shimshoni, R. Kimmel, and A. M. Bruckstein. Dialogue: Global shape from shading. *CVIU*, 64(1):188–189, 1996. [118]

[197] N. Sochen, R. Kimmel, and R. Malladi. From high energy physics to low level vision. Report LBNL 39243, LBNL, UC Berkeley, CA, Aug. 1996. Presented in ONR workshop, UCLA, Sept. 5, 1996. [144, 147, 150]

[198] N. Sochen, R. Kimmel, and R. Malladi. A general framework for low level vision. *IEEE Trans. on Image Processing*, 7(3):310–318, 1998. [55, 59, 144, 146, 147, 149, 150]

[199] N. Sochen and Y. Y. Zeevi. Images as manifolds embedded in a spatial feature non-Euclidean space. In *IEEE ICIP'98*, pp. 166–170, Chicago, IL, 1998. [151]

[200] G. A. Sod. *Numerical Methods in Fluid Dynamics*. Cambridge University Press, 1985. [65, 66]

[201] A. Steiner, R. Kimmel, and A. M. Bruckstein. Shape enhancement and exaggeration. *Graphical Models and Image Processing*, 60(2):112–124, 1998. [158, 183]

[202] W. S. Stiles. A modified Helmholtz line element in brightness-colour space. *Proc. Phys. Soc. (London)*, 58:41, 1946. [151]

[203] D. Terzopoulos. On matching deformable models to images. In *Topical meeting on machine vision, Technical Digest Series, Optical Society of America*, Vol. 12:160–163, 1987. [4, 11, 123]

[204] D. Terzopoulos, A. Witkin, and M. Kass. Constraints on deformable models: Recovering 3D shape and nonrigid motions. *Artificial Intelligence*, 36:91–123, 1988. [4, 11, 123]

[205] J. N. Tsitsiklis. Efficient algorithms for globally optimal trajectories. *IEEE Trans. on Automatic Control*, 40(9):1528–1538, 1995. [87, 134]

[206] M. Turk and A. Pentland. Eigenfaces for recognition. *J. Cognitive Neuroscience*, 3(1):71–86, 1991. [176]

[207] H. von Helmholtz. *Handbuch der Psychologischen Optik.* Voss, Hamburg, 1896. [150, 151]

[208] J. J. Vos and P. L. Walraven. An analytical desription of the line element in the zone-fluctuation model of colour vision II. The derivative of the line element. *Vision Research,* 12:1345–1365, 1972. [151]

[209] J. Weickert. Scale-space properties of nonlinear diffusion filtering with diffusion tensor. Report no. 110, Laboratory of Technomathematics, University of Kaiserslautern, Kaiserslautern, Germany, 1994. [155, 156, 158]

[210] J. Weickert. Multiscale texture enhancement. In *Computer Analysis of Images and Patterns; Lecture Notes in Computer Science,* Vol. 970:230–237, Springer, New York, 1995. [156, 158]

[211] J. Weickert. Coherence-enhancing diffusion of colour images. In *Proc. VII National Symposium on Pattern Rec. and Image Analysis,* Vol. 1:239–244, Barcelona, 1997. [145, 155, 156, 158]

[212] J. Weickert. *Anisotropic Diffusion in Image Processing.* Teubner Stuttgart, 1998. ISBN 3-519-02606-6. [133, 156, 158, 160]

[213] J. Weickert, B. M. ter Haar Romeny, and M. A. Viergever. Efficient and reliable scheme for nonlinear diffusion filtering. *IEEE Trans. on Image Processing,* 7(3):398–410, 1998. [133, 135]

[214] E. W. Weisstein. *The CRC Concise Encyclopedia of Mathematics.* CRC Press, Boca Raton, FL., http://www.astro.virginia.edu/~eww6n/math/, 1998. [28]

[215] R. Whitaker and G. Gerig. Vector-valued diffusion. In B. M. ter Haar Romeny, editor, *Geometric-Driven Diffusion in Computer Vision,* pp. 93–134. Kluwer Academic Publishers, The Netherlands, 1994. [145]

[216] A. P. Witkin. Scale space filtering. In *Proc. of the 8th Intl. Joint Conf. on Artificial Intelligence,* pp. 1019–1022, Karlsruhe, Germany, 1983. [56]

[217] G. Wyszecki and W. S. Stiles. *Color Science: Concepts and Methods, Qualitative Data and Formulae (2nd edition).* John Wiley & Sons, New York, 1982. [151]

[218] S. D. Yanowitz and A. M. Bruckstein. A new method for image segmentation. *Computer Vision, Graphics, and Image Processing,* 46:82–95, 1989. [145]

[219] A. Yezzi. Modified curvature motion for image smoothing and enhancement. *IEEE Trans. IP,* 7(3):345–352, 1998. [150]

[220] S. Zhu, T. Lee, and A. Yuille. Region competition: Unifying snakes, region growing, energy/bayes/mdl for multi-band image segmentation. In *Proc. ICCV'95,* pp. 416–423, Cambridge, June 1995. [186, 187]

[221] G. Zigelman, R. Kimmel, and N. Kiryati. Texture mapping using surface flattening via multi-dimensional scaling, visualization and computer graphics. *IEEE Trans. on Visualizationand Computer Graphics,* 9(2):198–207, 2002. [164]

Index

Printed by Books on Demand, Germany